SMITHSONIAN INSTITUTION PRESS • WASHINGTON, D. C.

The First 25 Years in Space

A SYMPOSIUM · EDITED BY ALLAN A. NEEDELL

© 1983 by the Smithsonian Institution.
All Rights Reserved

Printed in the United States
97 96 95 94 93 92 91 90 89
            5   4   3   2   1

Library of Congress Cataloging in Publication Data

Main entry under title:

The First 25 years in space.

   Proceedings of a symposium held Oct. 14, 1982 at the
National Academy of Sciences, and sponsored by the
Academy and the National Air and Space Museum.
   Supt. of Docs. no.: SI 1.2:Sp 1/2
   1. Astronautics—Congresses.   I. Needell, Allan A.
II. National Academy of Sciences (U.S.)   III. National
Air and Space Museum.   IV. Title: First twenty-five
years in space.
TL787.F5   1983   629.4   83-600210

ISBN 0-87474-668-X
ISBN 0-87474-713-9(pbk.)

British Library Cataloguing in Publication Data is
available.

Photo credits: Title page, pages 45, 83, 127, NASA
photos; page 1, National Air and Space Museum.

First time in paperback   1989

The paper used in this publication meets the minimum
requirements of the American National Standard for
Permanence of Paper for Printed Library Materials
Z39.48-1984.

Contents

FOREWORD   vii

ACKNOWLEDGMENTS   viii

INTRODUCTION TO THE 1989 EDITION
  ALLAN A. NEEDELL   ix

OPENING REMARKS
  FRANK PRESS   xii
  NOEL W. HINNERS   xiii

SECTION 1   THE MOTIVATIONS FOR SPACE ACTIVITY

Introduction   3
  JOHN LOGSDON
Motivations for the Space Program: Past and Future   6
  HARVEY BROOKS
Comments   27
  BERNARD SCHRIEVER
Comments   33
  AMITAI ETZIONI
Discussion   37

SECTION 2   THE PRACTICAL DIMENSIONS OF SPACE

Introduction   47
  WALTER SULLIVAN
The Practical Dimensions of Space   51
  SIMON RAMO
Comments   72
  ROGER CHEVALIER
Comments   76
  EDWIN MANSFIELD
Discussion   79

SECTION 3 SCIENCE AND SPACE

Introduction   85
  GERALD HOLTON
Science and Space   90
  FREEMAN J. DYSON
Comments   107
  HENDRIK VAN DE HULST
Comments   115
  GERALD J. WASSERBURG
Discussion   122

SECTION 4 CONCLUSION

Concluding Remarks   129
  PHILIP MORRISON

INDEX   149

Foreword

David Challinor
Assistant Secretary for Science
Smithsonian Institution

The National Air and Space Museum of the Smithsonian Institution has joined the National Academy of Sciences in this effort to examine the motivations, implications, and accomplishments of the first 25 years in space.

It is particularly fitting for the museum to be involved in this celebration because, according to our agreement with the National Aeronautics and Space Administration (NASA), the Smithsonian Institution receives all historically significant artifacts sent into space. Many are on display at the National Air and Space Museum, along with such authentic test and back-up equipment as the Lunar Module and the *Viking* Mars lander. Particularly notable is a replica of *Sputnik I*—a gift of the Soviet Academy—which is proudly displayed in the main entrance gallery with others "Milestones of Flight." It is the only artifact in that gallery not made in the United States. We do hope to acquire other space artifacts from abroad because space is truly international, exceeding by far the bounds of earthly sovereignty.

As charged in James Smithson's will, the task of the Smithsonian is to increase and diffuse knowledge among all mankind. This volume is the end product of a symposium held on October 14, 1982, at the National Academy of Sciences, whose close ties with the Smithsonian Institution began in the days of Joseph Henry, the first Secretary of the institution and one of the founders of the academy. That event and this volume fulfill Smithson's mandate well.

Acknowledgments

WALTER J. BOYNE
Director
National Air and Space Museum

I gratefully acknowledge the financial support of the Bendix Field Engineering Corporation, the Boeing Company, Fairchild Industries, the Space Systems Division of the General Electric Company, the Federal Systems Division of the International Business Machines Corporation, the Martin Marietta Corporation, the McDonnell Douglas Astronautics Company, the Government Systems Division of the Radio Corporation of America, Rockwell International Corporation, TRW, United Technologies Corporation, and Vought Corporation, who made the live symposium and this volume possible. I am very appreciative of their rapid response and their enthusiasm for the project.

I would also like to acknowledge the efforts of several people: Bruce Gregory and Jean Yates of the National Academy of Sciences, and Paul A. Hanle and Allan Needell of the National Air and Space Museum. They saw this project through from conception to completion.

Introduction to the 1989 Edition

ALLAN A. NEEDELL
Space Science and Exploration Department
National Air and Space Museum
Smithsonian Institution

The first assignment I received upon joining the staff of the National Air and Space Museum was to propose and organize an exhibit and public programs to commemorate the approaching anniversary of the first artificial earth satellites. This occurred in early 1981.

Ultimately, a decision was made to create a temporary exhibit to open on July 1, 1982, to commemorate the 25th anniversary of the formal beginning of the International Geophysical Year (IGY). The IGY was actually an internationally agreed upon 18-month period of cooperative scientific investigations into wide areas of scientific study. It was in conjunction with the IGY that the Soviet Union and the United States undertook to launch instrumented "earth-circling" spacecraft to report on the upper reaches of the ionosphere and near earth space. *Sputnik I* was launched into orbit just over three months into the IGY; the United States's *Vanguard* satellites were scheduled for launch in 1958.

Another decision was to create a permanent record of contemporary informed assessments of the history and future of space exploration. A public symposium and this volume were the results. Clearly the most dramatic of the pending quarter-century anniversaries was that of the *Sputnik* launch. A major event like that presented a special opportunity to convene a symposium and to assess the technological trends and the political and other forces that had buffeted the space programs of this and of other countries.

As is pointed out several places in this volume, retrospective and prospective assessments of the space age are likely to say at least as much about the period in which they are uttered as about the periods under discussion. But our judgment was that such assessments, coming during the early stages of the Reagan administration, might well prove to be of more than passing interest. From the vantage point of 1989, on the occasion of the beginning of a new administration in Washington, I am convinced this was correct.

The selection of speakers, a task for which I received help from John Logsdon—the introducer of section 1 following—proved to have been fortunate. In almost every case, the perspective of the symposium participants has come to seem even more significant with the passing of time. For the most part, the issues that were raised and discussed—to be sure in the context of 1982—have become even more focused and more pressing in the years that have followed. Just a few examples:

Harvey Brooks and several others raised the issue of whether the high profile manned space flight effort inappropriately diverted resources away from exploration and science or whether the national commitment to manned programs like the lunar landings raised the level of support for exploration and science above the levels they would command on their own. Brooks also assessed the costs and benefits of maintaining essentially separate military and civilian space programs. Brooks and others commented on whether competition with our Western allies in the fields of space science and applications would ultimately serve the consumer and society by encouraging innovation, or cost society as a result of the wasteful duplication of large-scale technological efforts.

General Bernard Schriever presented an important and unique perspective on the question of the militarization of space in the days before the dramatic March 1983 announcement of President Reagan's Strategic Defense Initiative. Simon Ramo provided interesting speculations about the economics of various proposed space-based industries. He also provided an impassioned plea (still unheeded) for the establishment of rational and workable assignment of government and private roles and organizations for important space applications, especially an international space-based system for air navigation and traffic control. And, in the "Science and Space" section, Freeman J. Dyson in his uniquely provocative manner raised the issue of big versus small technology emphasis in the space science programs of this and other countries.

Finally, Philip Morrison discussed timeless issues of the relevance of advanced technology to the lives and welfare of the vast majority of people ,inhabiting our planet. He concludes with a sobering discussion of the potential of our surviving as species with or perhaps in spite of our technological triumphs.

In my opinion, this symposium volume has more than retained its value. And I strongly suspect that as time goes on the availability of this snapshot of thoughtful, informed consideration of important issues by characters central to the first quarter century of space flight will continue to serve those who seek to understand the crucial relations of science, technology, and society in the twentieth century.

Opening Remarks

FRANK PRESS
President
National Academy of Sciences

It is difficult to believe that 25 years have passed since the launch of *Sputnik I*. That event triggered a sense of great inadequacy in the United States in the area of science and technology education. The country reacted by pouring vast sums of money into buildings, laboratories, fellowships, research and development programs, and revised school curricula. Twelve years later the *Apollo* spacecraft landed on the moon's Sea of Tranquility. Then the deflation began. How many times have we heard "we need another sputnik" to galvanize the nation to do this or that? Just recently a cabinet officer said, "We need another Sputnik to trigger a vast reorganization and improvement in the nation's precollege science and math education programs."

Of course, the real question is: Why do we again need national improvements in so many areas today, 25 years after *Sputnik*, and what event will trigger a national response to some of these issues and problems? I do not believe it will be another sputnik, but it may be the new technological competition between the industrial allies.

The past 25 years in space have largely been a story of competition and cooperation between the United States and the Soviet Union. The next 25 years will be more difficult to characterize, with the entry of the Europeans and Japan into space exploration and space applications. I think the future will see a sort of dichotomy; we may see growing competition commercially and growing cooperation scientifically in space programs. We will probably see major ventures that are jointly planned, constructed, and financed by many nations in the area of scientific research in space. And perhaps, sometime in the next century, a world space program will lead to major missions, perhaps manned, to one of the planets.

There was a direct connection between the initiation of the space age, the National Academy of Sciences, and the International Geophysical Year (IGY). The launch of the early satellites was one activity among many carried out during the International Geophysical Year, which had its origins at the home of James Van Allen in Silver Spring, Maryland. Academicians Lloyd Berkner and Sidney Chapman were present and thousands of scientists ultimately participated, not only in the IGY but in subsequent programs like the International Hydrological Decade and so on. The academy played a major role in organizing and implementing U.S. participation in the IGY and these other programs.

This symposium, then, addresses the motivation for the exploration of space, assesses the accomplishments of the past 25 years, and evaluates their impacts on society. There are many stimulating and challenging presentations and discussions.

It is a day of history, and it is a day of looking forward to what the future will bring.

NOEL W. HINNERS
Director
Goddard Space Flight Center

In one previous incarnation I was the director of the National Air and Space Museum. Indeed, it was at that time that we began working on the development of this symposium. Now, with a sense of wary anticipation, I survey the results. My feelings are typical for this type of endeavor: Do the authors feel that they have produced fair value for the time expended in writing their papers? Will the audience perceive that they have participated in something more than an exercise of self-congratulation?

To arrive at a fair assessment of these questions, one must know how this symposium was conceived. First, the occasion of a 25th anniversary for the space age presented a convenient excuse (if you will) for taking stock of the causes, initiation, development, and rewards of space exploration. Twenty-five years is, I believe, a sufficiently long time for the beginnings to have ripened enough to warrant a decent historical evaluation. On the other hand, the time is sufficiently short so that many of the pioneers of the effort are still available to lend firsthand data and impressions to the account.

The intent is also to celebrate an amazing accomplishment of humanity, brought about by a combination of political, scientific, and technological resolve. The lasting beneficial fruits of the first 25 years of space exploration, I think, reside in the contribution to our store of basic knowledge and to an uplifting of the human spirit. The two travel hand in hand, and both are represented in this symposium.

It was our desire to temper the retrospective look with constructive criticism, lest we not learn from experience, and I would conjecture that perhaps the most significant lesson will be that the details of history are unlikely to recur. Thus, rather than yearn nostalgically for the good old days, let us resolve to develop new ways to give meaning to a 50th-anniversary assembly here by our descendants in the year 2007. Its cosponsors should also be the National Academy of Sciences, catalyst and advocate for unfettered, vigorous pursuit of free scientific inquiry, and the National Air and Space Museum, chronicler, preserver, and interpreter of the historical context in which space exploration occurs.

SECTION 1

The Motivations for Space Activity

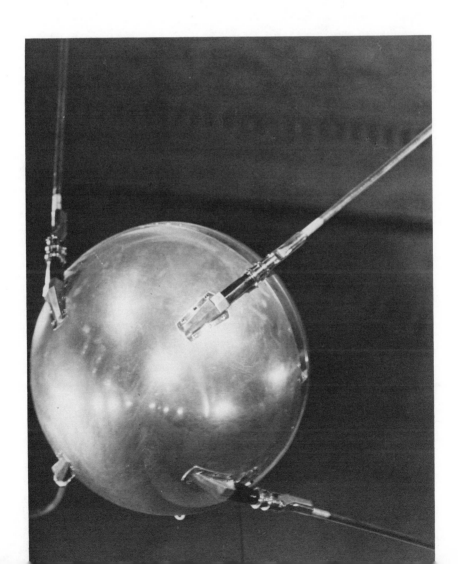

Introduction

JOHN LOGSDON
Graduate Program in Science, Technology, and Public Policy
George Washington University

One of the immediate reactions in the United States to the launch of *Sputnik I* was to create a President's Science Advisory Committee (PSAC). In the very first sentence of its first report, in March 1958, the committee asked the question that concerns us today: "What are the principal reasons for undertaking a national space program?" In the intervening quarter-century we have come a long way; we no longer need a PSAC to answer the next question raised in its report: "Why do satellites stay up?" But the first question remains unresolved to a large degree. This initial section of the symposium is part of a continuing reexamination of why the United States and other countries and private interests of the world are and should be investing large sums of money in space activities. In this country, I would argue, there has never been anything approximating a consensus on what we want and expect from space, except possibly for that brief period when we entered the space race in competition with the Soviet Union, with a full commitment to winning.

In 1958 the PSAC identified the following four factors, which gave "importance, urgency, and inevitability to the advancement of space technology":

- the compelling urge of man to explore and to discover;
- the defense objective, i.e., if space is to be used for military purposes, we must be prepared to use space to defend ourselves;
- national prestige; and
- new opportunities for scientific observation and experiment.

3

Conspicuously missing from this early listing was the potential of applying space technology to produce societal benefits and commercial payoffs. Even with this addition, today's list of motivations for space activity would differ little from that of a quarter of a century ago:

- scientific discovery;
- national security;
- national image; and
- beneficial applications.

What is controversial—then and now—is what priority should be given space activity among other uses of advanced technology and, within the realm of space efforts, which motivations should carry the greatest weight. This symposium should help us think through these continuing issues.

Science, security, and society have been three of the dominant motivations for human activities in space. Thus it is quite appropriate that this first section consists of contributions from a scientist, a soldier, and a sociologist, each of whom has an extremely broad perspective on the interactions among science, technology, and policy that condition so much of contemporary human affairs.

The focal point of the section is a most thoughtful paper by Harvey Brooks on the motivations of the U.S. space program. Harvey Brooks has been omnipresent in the U.S. research system over the past quarter-century, and the country is better off for his involvement. He is a physicist by training, and most of his professional career has been at Harvard, where he was the Gordon McKay Professor of Applied Physics from 1950 to 1975. No advisory committee is complete without his involvement somewhere along the line. He was a member of the PSAC during its most exciting years, 1959–1964, and he has been active in all arenas of international science policy. Brooks is a member of the National Academies of Sciences and Engineering, the American Philosophical Society, and the American Academy of Arts and Sciences, of which he is past president.

Bernard Schriever and Amitai Etzioni provide comments on the Brooks paper. Bernard Schriever retired from the U.S. Air Force in 1966 with the rank of General. As the commander of our ICBM program, he was intensely involved with the development of the U.S. strategic missile capability in the 1950s, and he headed the Air Force Systems Command from 1959 to 1966. In both positions he was at the center of the ongoing debate over the national security uses of space, and he has continued to speak out on this issue in recent years as he pursues a second career as a top-level management consultant.

After a long career at Columbia University, Amitai Etzioni two years ago become the first University Professor at my home institution, George Washington University. He is hard to categorize; a sociologist by training, he identifies himself as a political sociologist or macrosociologist. But those who are familiar with his work know that Etzioni's driving curiosity and range of intellect have led him to explore every facet of contemporary life. Perhaps the title of his most recent book suggests the breadth of his concern: *An Immodest Agenda*. A recent survey found him the most cited academic analyst of public policy over the past quarter-century.

Etzioni first turned his attention to the space program in the early 1960s, publishing a rather critical analysis called *The Moondoggle*. His comments here will show if two decades have tempered his evaluation.

When I was writing of the decision to send Americans to the moon, I came to a conclusion that seems relevant to the subject of this section. Like all major policy choices, President Kennedy's Apollo commitment was a product of a specific time and circumstances. Through that commitment, "the politics of the moment became linked with the dreams of centuries and aspirations of the nation." What appears lacking in space policy today is a link between continuing national motivations for undertaking activities in space and the priorities of the time. Perhaps these contributions will help us decide whether this link should be repaired.

Motivations for the Space Program: Past and Future

HARVEY BROOKS
Division of Applied Sciences
Harvard University

INTRODUCTION AND EARLY HISTORY

Man's escape from earth to explore space had appealed to his imagination long before it became possible. Unlike most of the major technological revolutions of the 20th century, the adventure in space was essentially a 19th-century dream. That dream was based largely on technologies understood in general terms at the time, although its practical realization required 20th-century developments in many ancillary technologies (materials, electronics, radar, radio communication). When Jules Verne imagined a voyage to the moon in 1865, he wrote about it with considerable technical realism. H. G. Wells also envisioned space travel. Indeed, 19th-century fiction reveals much more foresight about space travel than about air travel, which actually required a good deal more really new science than did space.[1]

The motivations of the pioneers who created space technology were different from those of the societies that began to support it on a large scale. The vision of the pioneers was always man-in-space and man's exploration of the solar system. If this had to be achieved under the guise of military necessity, the early space engineers were prepared to suppress their real priorities in order to convince governments to support them. In the beginning at least, the military applications were probably largely tongue-in-cheek. Even Wernher von Braun and his team in Germany appear to have been interested

6

mainly in space exploration, and saw the V-2 program as the most politically feasible first step toward the realization of that dream.[2] When the early U.S. space program had bogged down, von Braun, by then an American, was eager to move his team into the civilian space effort.

Space, like atomic energy, was a technology that started in the military, but eventually reached a point where the development priorities for military and civilian applications began to diverge. In the USA the first serious thinking about the possibility of orbiting satellites apparently occurred with the creation of the Rand Corporation in 1946. A Rand report of that year foresaw with uncanny accuracy the worldwide political effects that might occur if a relatively backward power such as the Soviet Union were to succeed in orbiting a satellite before the United States did. The report urged the U.S. Air Force to mount an effort to foreclose this possibility.[3]

However, at that time the political significance of spectacular high technology per se was not appreciated, and rocket technology continued to be viewed in terms of demonstrable near-term military applications. According to the Rand report, the satellite was a feasible device but not a military weapon. Because it was not a weapon with a specific military requirement, it was impossible in the political climate of the 1940s to make funds available for its development. In vain did the Rand authors plead that, "in making the decision as to whether or not to undertake construction of such a craft now, it is not inappropriate to view our present situation as similar to that in airplanes prior to the flight of the Wright brothers. We can see no more clearly now all of the utility and implication of spaceships than the Wright brothers could see flights of B-29's bombing Japan and air transports circling the globe."[4] They concluded: "It is therefore recommended that the satellite be considered not as an academic study but as a project which merits planning and establishing of a priority in the research program of the Army Air Force."

No near-term military use for satellites was foreseen in the 1950s, and development, though not dropped, was given low priority. The idea was treated as a scientific toy, intriguing and appealing to the imagination, but not important enough to interfere with more urgent military developments. Thus the U.S. satellite program for the IGY was not permitted to use a military booster for fear this would detract from the more urgent ICBM program.[5] Similarly, 20 years passed before engineers and politicians gave any serious condition to Arthur Clarke's proposal for a geostationary communications satellite.[6]

THE SPUTNIK SHOCK

The Soviet launching of an orbiting satellite in 1957, using a military booster, completely transformed the climate of leadership opinion not only in the United States but around the world, amply confirming the predictions of the Rand authors. Not since the explosion of an atomic bomb over Hiroshima had a technological event had such immediate and far-reaching political fallout. *Sputnik I* was an extraordinary shock to American and world elite opinion. We realize now its significance was enormously exaggerated, not just by the media, but by a good many who (in the wisdom of hindsight) should have known better. In fact, its technological and scientific significance was overestimated; its political and psychological effect was not. Typical was the comment of the elder statesman Bernard Baruch: "America is worried. It should be. We have been set back severely not only in matters of defense and security, but in the contest for the support and confidence of the peoples of the world."[7] According to R. Cargill Hall: "Soviet satellite successes more than fulfilled the overall political/psychological expectations [of the Rand report] in reverse. It precipitated Congressional investigations into United States missile and satellite programs. Defense Department reorganizations were recommended. Bitter accusations on the 'missile gap' were traded among top military and administration personalities, followed by a number of resignations of key personnel in the military services." However, the political effect would probably have been far more lasting if America had not risen so promptly and visibly to the challenge.

Sputnik I and Soviet space achievements of the subsequent several years created a crisis of confidence in American power and moral leadership. The event was taken by elite opinion everywhere to demonstrate a laxness and misplaced complacency in this country's values and priorities in fields ranging from education to consumer tastes. The tailfins just appearing on the latest models of American cars became the visible symbols of U.S. decadence and the misallocation of our engineering talents. In quick succession there followed the National Defense Education Act, designed to identify and recruit talent for science and engineering as well as other intellectual pursuits; the curriculum reform movement for science in the schools; the transfer of the President's Science Advisory Committee into the White House, with a full-time science advisor reporting directly to the President; and a sharp increase in funding for basic science across the whole spectrum of disciplines. The 1960 political campaign saw an unprecedented emergence of science and technology issues into the political rhetoric of the candidates. All of this finally culminated in 1961 in the Kennedy commitment to land men on the moon and return them safely to earth by the end of the decade, a commitment which passed the Congress with only one dissenting vote and which was universally applauded by the press and public and most of educated public opinion around the world.[8]

The scientific community viewed all this with some ambivalence. It was nice to be loved and famous, but all but a few space scientists feared that the new national priorities would devour resources that should be spent on more valuable science rather than technological spectaculars. The ambivalence of the scientific community surfaced in a disagreement between Kennedy and his science advisory committee, a disagreement that received little attention at the time. The committee felt that the investment in the Apollo program could not be justified in terms of the scientific results likely to be obtained and Kennedy was aware of this. He told his science advisor, however, that Apollo was not a scientific but a political project, and they agreed to avoid a confrontation by not asking the PSAC's advice.[9]

Both Kennedy and the PSAC were probably right. If government expenditures for research and development could be viewed as a fixed sum of money to be allocated optimally among scientific projects, clearly no scientist, probably not even a space scientist, would have advised spending such a large proportion of those funds on Apollo. But the technical community has since learned that R&D money is not fungible, especially under crisis conditions. The effects of crises on science have been much more to increase the size of the total science pie rather than to reallocate relatively fixed funds to new priorities. In retrospect, it seems fairly clear that the Apollo program, far from competing with federal support of science, pulled the level of science support up with it.

Furthermore, Kennedy was probably correct that circumstances demanded a highly visible and easily understandable demonstration of American technological prowess to offset the psychological damage not only of *Sputnik*, but also of the Bay of Pigs fiasco and the confrontation with the Soviets over Berlin. Apollo provided the means for such a demonstration without directly threatening the USSR or raising public fears of a military confrontation. It was like a challenge between the champions of two medieval armies, the race for the moon serving as a partial surrogate for more threatening forms of competition.

In fact, the United States responded to the perceived challenge of *Sputnik* in both military and technological terms. The early 1960s saw the growth of a host of military programs, from Polaris to Minuteman, and defense spending reached a peak as a fraction of the gross national product not attained before or since. But the visibility of the Apollo competition probably, if anything, diverted attention from the military competition and helped to avoid an even greater militarization of American society than might have occurred.

In his book, *The Tools of Empire*, D. R. Headrick has shown how 19th-century European, particularly British, technology not only provided the instrumentalities for colonial penetration of much of what we now call the Third World, but also conferred on the colonizing societies a sense of moral superiority derived from their technological superiority.[10] They came to feel that their technological prowess derived from their moral superiority, and this armed them morally to undertake great hardships and risks in bringing "civilization" to the less developed world. There was probably also an important moral dimension to the Apollo achievement which was of the same character, but which probably also contributed to the moral *hubris* that led to the embroilment in Vietnam by the end of decade. An element of this same moral dimension exists today in the feeling evident throughout the world that the mystique of high technology confers political legitimacy on a nation vis-à-vis other nations; thus, every government feels constrained to demonstrate its mastery of certain technologies without becoming dependent on others for them. It is not too far-fetched to suggest that in most countries the mastery of high technology is perceived, at least in part, as a demonstration of civic virtue.

What is remarkable—in the perspective of the subsequent history of more faltering American political commitments to energy independence, the elimination of poverty, and cleaning up the environment—is the fact that the Apollo commitment was sustained with so little change over a period of more than eight years of political turbulence and a "sea change" of national mood. This is largely due to the fact that Apollo was an easier goal to accomplish, but it also owes something to the political and managerial skill of then NASA Administrator Jim Webb. He was able to insulate the technological integrity of the program from political faction and sniping. This was not the case with any other American technological commitment outside the defense field, which enjoyed the advantage of being partially screened from public scrutiny by the cloak of security classification.

THE PRINCIPLES OF THE AMERICAN SPACE PROGRAM

It is surprising that the underlying principles of the program as first formulated in the 1958 Space Act have remained nearly constant to this day. They include basically the following features:[11]

• the separation of military and civilian space activities;
• a high degree of openness in the program, a strong emphasis on public information, and a willingness to expose mistakes as well as successes;
• a strong international orientation (The Act declares that the USA is to exploit space for "peaceful purposes for the benefit of all mankind" and it mandates "cooperation by the U.S. with other nations and groups of nations." However, this cooperation probably implicitly presumes "preservation of the role of the United States as a leader in aeronautical and space science and technology.");[12] and

• a government-directed program managed by civil service laboratories, but with more than 80 percent of the activity delegated to a host of contractors and subcontractors. (This follows the mandate of the 1958 Act to have "the most effective utilization of the scientific and engineering resources of the United States," and follows the pattern originally set by the Office of Scientific Research and Development (OSRD) during the Second World War and later copied in the legislation establishing the Atomic Energy Commission, the National Science Foundation, and the National Institutes of Health.)

Such a pattern of devolution of management was particularly difficult in a program as focused and integrated as Apollo. It required the orchestration of hundreds of quasi-autonomous, private organizations, under contract not only in connection with development, design, and construction, but also for the conduct of actual flight operations from Houston and Cape Canaveral. The resulting management structure demanded a unique blend of hierarchy and collegiality, which has to some extent become the pattern for other high-technology undertakings and is perhaps the most unique contribution of the space program to the art of management.[13]

TENSIONS AND CONTRADICTIONS IN THE ORIGINAL PROGRAM

The national debate that led to the 1958 Space Act and subsequently to the structure of the Apollo program revealed a number of tensions and contradictions, which have remained a persistent thread in the subsequent evolution of the U.S. space program. They are still very much in the forefront of the national discussion about its future. I list them here as background for a subsequent analysis of how the same issues are shaping debate over the future of the U.S. space effort.

1. *Manned vs. unmanned space projects.* Even as early as 1961 many scientists believed that man's presence in space was unnecessary for the achievement of any rational objective in a national space effort. They felt that all the scientific objectives of the program could be

attained at much lower cost, and at substantially less political risk, by using unmanned vehicles with sophisticated automation and remote control through communications links to earth. It was conceded that this might entail a considerably longer program, but one with less technical risk. Others argued that the presence of human judgment on the spot outweighed the risks of a much more embarrassing fiasco should lives be lost in a space mission. In the end it was probably the political imperatives of beating the Russians on a constrained time schedule that won out, though the romantic dream of the science fiction writers undoubtedly influenced this preference. Throughout the debate there was also the persistent fear of the scientific community that the inevitable budgetary overruns in a large program with so many technical risks and uncertainties would end by crowding out more scientifically rewarding unmanned space experiments—something which, to some extent, probably did happen.

2. *A civilian vs. a military space agency.* The question was debated as to whether it was realistic to try to separate the civilian and military space efforts when they had so many underlying technologies in common. Would a civilian space agency lead inevitably to wasteful duplication of effort and draw critical resources and talent away from what many saw as a more urgent military effort? Given the overlap of technologies, what would have to be under security classification in the civilian agency in order to avoid compromising U.S. military capabilities in space? These issues arose acutely in the debate between proponents of earth orbit and lunar orbit approaches to the moon landing. A strong argument of those favoring the earth orbit approach was that the resulting technological capability more readily extrapolated to military applications. Although this issue quieted down with the triumph of the lunar orbit approach, it has once again come to the fore with the advent of the shuttle and the perceived increased attractiveness of carrying military and civilian payloads on a single platform (not just the shuttle).

3. *International orientation of the program.* Of all the program's features this was probably the least questioned at the time, in part because the U.S. lead in space technology seemed unassailable, except by the Soviets, and in part because politically the U.S. needed measures to deemphasize its perceived role as being interested largely

in military power. Since the goals of the new U.S. space program were avowedly political rather than military, it made sense to stress the international and cooperative aspects consistent with the current American approach to the political rivalry with the Soviet Union. That approach had begun with the Marshall Plan and extended through the Atoms-for-Peace initiative and most recently the Eisenhower "open skies" proposal. In each of these examples the United States had gone out of its way to stress the nondiscriminatory nature of its offer to share its technological capabilities with others with minimal emphasis on political preconditions. For example, the offer of Marshall aid originally included the Eastern block, but the USSR rejected it. Atoms-for-Peace was offered to all comers, and the "open skies" proposal was to embrace all nations. This approach was more consistent with the concept of leadership embodied in the Space Act, rather than with hegemony, which was advocated by enthusiasts of the military emphasis in space.

4. *NASA as an R&D rather than an operating agency.* This problem was latent at the beginning. Few of the framers of the 1958 Space Act envisioned a day when space would become the site of routine operations; almost nobody anticipated how quickly this would happen. Therefore, the question of who should be responsible for operational or commercial services based on space technology seemed largely theoretical and academic and received little attention in the debates leading to passage of the Space Act. The only NASA responsibility that could be described as operational involved launching and tracking, but as long as this served primarily experimental and research functions it could be viewed as a support function rather as an operational responsibility. Even Apollo, though operational in a sense, was too experimental to present problems: There were no users outside of NASA itself, and the hardware was largely handmade rather than mass-produced.

Nevertheless, the question of operations responsibility quickly became an issue as, first, weather observation, then communications, and most recently remote sensing and navigation from space platforms became practical possibilities for routine service operations. Each offered an alternate means of providing a routine service that was already being furnished by other government or private agencies using more traditional technologies. Thus, the transition to service involved competitive jurisdictions between NASA and existing government or private organizations.

Because its responsibility was to generate technology, not to use it, NASA tended to give rather low priority to the requirements of potential users. It remained a "technology push" rather than a "market pull" organization. Unkind critics described NASA's function as generating "solutions looking for problems," which is, of course, a problem that exists within the developmental part of any high-technology enterprise. Problems of transition to commercial application are always difficult, in part because improvements in technology take place more rapidly than revenue from operations can recover the costs of development. In the case of NASA, the transition problem has been exacerbated by the circumscription of its mandate in the original legislation, and by the fact that the appropriations structure of the federal government makes no distinction between investments that produce revenue or utility streams over time and straight expenditures. The same problem has plagued the government's nuclear power program, particularly in the provision of enrichment services and radioactive waste management. In a sense, some of the problems faced by the space program have been the product of its own successes, which were not sufficiently well planned for.

All of the issues outlined above apply in the context of the future of the space program. We will consider them in sequence.

MANNED VS. UNMANNED SPACE PROGRAM

Man-in-space continues to have a great popular appeal, as evidenced by the revival of public enthusiasm for the space program that accompanied the first successful flights of the shuttle.[14] Manned exploration of the near planets remains a glamorous and appealing goal, and serious scientists continue to talk of space colonies and manufacturing on the moon. When the future of the space program was being debated in the immediate post-Apollo period, the nation gradually became committed to the shuttle as the next major step in manned space technology. But, at the insistence of the Vice President, as chairman of the Space Task Group of 1969, the option of future manned planetary exploration was kept alive, though deferred.[15]

However, in the 20 years since the first commitment to Apollo, tremendous progress in electronics, sensors, automation, and communications has transformed the comparative advantage of men and machines in space so that more people than earlier can question whether there is any legitimate practical function for man, apart from the ever-present elements of popular appeal and political posturing. Does man have any incontestable advantages in space, rather than on the ground at the end of a sophisticated communications link?

Man does have one advantage: on location in space he can be part of a feedback loop with a much shorter time delay than if he were on earth. Although one can envision ever-increasing sophistication of both instrumental sensing and manipulation on, say, a planet, with man's judgmental capacities being injected at the earth segment of the loop, there is a communications delay set by the velocity of light that cannot be compensated for by even the most elaborate signal processing of the time-varying data that come back from the planet. Events on the planet that cannot be extrapolated from recent history over the duration of two-way communications cannot be handled by a human operator on the ground. The question is whether such discontinuous events are sufficiently frequent or important in the exploration of the planets to constitute a significant limitation on remote exploration possibilities. This is a difficult question.

An important argument favoring the unmanned approach is that the technology is more likely to be adaptable to subsequent beneficial terrestrial applications. What can be learned from the challenge of designing remote manipulations and measurements in space is in principle much more directly applicable to such operations as automated or remotely operated factories on earth. One can even imagine automatic manufacturing systems capable of replicating themselves remotely in new locations. Indeed, this has already been suggested in connection with the possibility of building remote industrial operations in the deep oceans, at great depths in the earth, or in situations inhospitable to man such as high-radiation environments, fires, or poisonous atmospheres. By contrast, the technology necessary to support man in the hostile environment of space is much more specialized and less obviously or easily extrapolated to terrestrial tasks. Of course, there are incidental things to be learned from observing human responses under conditions of weightlessness, but the benefits are relatively small in relation to the cost of developing and implementing the technology.

The problem of budgetary overruns in costly manned programs disrupting planning and program continuity for unmanned science and applications programs, which surfaced during the Apollo era, has become even more acute with the shuttle. This is because budgetary stringencies are much more severe than they were in the 1960s. Moreover, this situation seems unlikely to change in the foreseeable future. It arises from the competition of increasingly expensive entitlement programs, programs whose growth—because of the political forces present in all industrial democracies—can at best be slowed but not reversed.

The issue has more severe practical consequences now, because Europe and Japan are developing competitive unmanned space capabilities, particularly in the space applications area, with total budgets less than one-fifth of those being devoted to space by the United States. Perhaps the USA is undermining its own long-range capacity for commercial competition in the provision of space-based services by allowing its science and applications programs to be disrupted by budgetary "crunches." Such disruptions are not the fault of the programs, but the result of fiscal restraints combined with the massiveness and inflexibility of manned programs. The problem is not just the size of manned programs but their "lumpiness"—the fact that they form single integrated packages, which cannot be constrained without serious risk to the technical success of the program, no matter how much the original cost may have been underestimated. Would manned space efforts continue to enjoy such high priority in the United States in the absence of competition with the Russians? If the USA were to devote the same effort to applications and sciences (as well as to basic aeronautical technology, which is also a NASA responsibility), the long-term benefits to its competitiveness in high technology might be greatly enhanced. Perhaps

Japanese competitiveness with the United States in expendable boosters, communications, and remote sensing satellites is a consequence
of the inflexibilities in the U.S. space program caused by internal
competition for resources associated with the space shuttle.[16]

There is no clear, long-term solution to this problem. Budgetary
competition within NASA is not a law of nature, and there is no
inherent reason why shuttle overruns should compete with other
NASA programs rather than with other government programs of
lower priority—no reason, that is, except long habit and bureaucratic
tradition. And it is likely to continue to spark intense debate, especially as financial resources for space activities become increasingly
constrained. Moreover, the politicians are likely to take the side of
the manned space activities, as in 1961, whereas most, though by
no means all, of the technical community is likely to favor the machines.[17]

The issue for the future is not so much the use of the now-available
shuttle technology to enable men to service, repair, or replace equipment in earth orbit. That technology is here and will undoubtedly
improve and become more cost-effective and reliable with time. Rather,
the question is whether there should be any further development of
manned capabilities, e.g., for deployment in geostationary orbit, or
ultimately for planetary exploration, or for permanent space stations.
Such developments incur future opportunity costs against automated
alternatives. Perhaps the development of manned capabilities should
be considered only after the exploitation of particular space sites
through automated and instrumented means appears to be reaching
a point of diminishing returns, which the introduction of man may
circumvent. In other words, manned space activities might be driven
by market pull generated from unmanned activities, rather than by
the technology push of what is within the current state-of-the-art,
as it clearly was in the Apollo case and to a partial extent in the case
of the shuttle. Such a scenario would assume that the incidental
political benefits of manned activities are too transitory to warrant
the investment, unless they can be unequivocally justified by technical requirements after unmanned alternatives have been thoroughly assessed.

CIVIL VS. MILITARY SPACE PROGRAMS

In my opinion, the separation of civil and military space activities has served the country well so far. Such a separation may be more difficult to maintain in the future, although I believe it is worth trying to do so. What will make the separation more difficult is the combination of budgetary pressures with the perceived economies of simultaneously using a single space platform or vehicle for a variety of different missions. Hence, the issue is not only civil vs. military but that of multimission platforms and vehicles in general. There is a question of whether the apparent economies of combining missions are real when the "transaction costs" arising from managerial complexity, schedule slippages among separate missions, and technical compromises necessary for satisfying multiple-mission requirements are fully considered. The emergence of foreign competition in launch services vis-à-vis the shuttle may be an indication of competitive problems, which will appear with increasing frequency if the multipurpose mission approach is pursued too exclusively.

In addition, the marriage of military and civilian missions in the same vehicle or platform introduces further complications. There are likely to be continual controversies over classified information and the access of the managers of civilian payloads to mission information they feel is necessary for effective management of their particular package. Where secrecy is necessary there is likely to be friction, especially when foreign payloads are involved.

Efforts of the United States to combine military and civilian payloads are likely to stimulate foreign countries to develop independent capabilities. The combination with military programs will be seen as lowering assurance of available flight opportunities, and nations unsympathetic to U.S. security objectives or foreign policy will see themselves as compromised by taking advantage of partially military space vehicles. Blending of civil and military missions will almost certainly undermine the principle of nondiscriminatory availability of American technological capabilities, which was implicit in the original concept of the 1958 Space Act.

The development of independent foreign space capabilities has both benefits and costs. The benefits arise from breaking the U.S.

monopoly on certain capabilities and thus stimulating a more in-novative and accommodating approach on the part of U.S. space vehicle managers. The consumers of space services, both American and foreign, would probably benefit from the more competitive sit-uation. Against this must be offset the economic costs of unnecessary duplication of development effort and investments. It is not clear which types of space services are best treated as "natural monopo-lies" and thus provided cooperatively, and which are more naturally competitive. In either case, it would be ironic if U.S. efforts to save budgetary costs by combining military and civilian space activities resulted in the further stimulation of foreign competition, even if that might ultimately benefit the potential users of space services. Such foreign competition is clearly an unintended consequence of present U.S. policy, and probably inconsistent with its desire to prevent the acquisition of militarily significant space capabilities by other countries. If a more competitive environment in space is in-trinsically desirable, the militarization of the U.S. space program is an awkward way to effect it.

The long-term political liabilities from intermingling military and civilian activities in single space missions are difficult to define, but they should be given more weight than they have recently in the United States.[18] The transfer of technology between military and civilian activities, on the other hand, is probably beneficial to both and does not carry such political liabilities. In particular, the adap-tation of military technologies to civilian applications, as in com-munications (utilization of the 20/30 MH band, for example), remote sensing, and navigation, can improve the image of the overall space program while expediting valuable civilian developments. Transfer in the other direction (from civilian to military) may be less probable. It is also less likely to be visible, so that it will raise fewer political problems.

The political liability stemming from combined military-civilian missions will also depend on the particular military application that is contemplated and its perceived legitimacy under existing inter-national understandings. For example, given recognition under treaty of the legitimacy of "unilateral means of verification," the launching of intelligence-gathering satellites from the shuttle or other platforms would probably not raise serious problems. On the other hand, the use of the shuttle or other platforms for experiments with laser or particle beam devices for ABM or anti-satellite purposes might be viewed as contrary to the spirit, if not the letter, of existing international agreements. Similarly, placing in orbit civilian navigation satellites capable of providing ships with the precise information necessary to launch silo-killing missiles would raise serious problems.

Any defense experiments on an otherwise unclassified mission might raise exaggerated suspicions as to the type and amount of military work being done under the guise of "peaceful" space activity. There is a considerable political advantage in being uncompromisingly open, without even minor exceptions. One can only point to the permanent damage to U.S. and world oceanographic research done by the revelation of the use of oceanography as a "cover" for intelligence activity in the *Pueblo* incident. Even American scientists are likely to have qualms about "piggy-backing" their experiments on space missions that are widely suspected, justifiably or not, to have a primarily military purpose, especially if this purpose is concealed or suspected of being concealed.

Moreover, the United States probably has a basic interest in avoiding enlargement of the domain of potential military conflict in space because it has more to lose than to gain from such enlargement. In the past the USA has evaluated proposed new weapons systems under the assumption that it has them and the other side does not. Although this situation may exist for a year or two after the introduction of a new weapon, it is probably very temporary, since the other side is likely to match new capabilities quickly.[19]

It is therefore more reasonable to assess new weapons proposals on the assumption that the resulting capability will be available to both sides. Under these circumstances the United States may be at a relative disadvantage, as happened, for example, with MIRV technology; or at best both sides may end up less secure than if neither had acquired the capability in the first place. It is hard to imagine any new space-based military capability that, if available to both sides, would still be advantageous to the United States. The only exception appears to be unilateral intelligence-gathering capabilities, which are relatively more advantageous to the USA because of the secretiveness of its adversaries.[20]

INTERNATIONAL COOPERATION AND TECHNOLOGY SHARING VS. COMPETITION

Its willingness to make its space capabilities available to other nations on a largely nondiscriminatory basis has been of great political benefit to the United States. There is, perhaps, more doubt as to the economic benefits of this policy, but on the whole cooperation in space has been a positive-sum game in which all have reaped benefits. Thus, the sharing of space capabilities probably remains advantageous to the United States both politically and economically, provided the country is prepared to continue investing enough to maintain at least a modest technological lead over its potential space service competitors. On the other hand, if potential competitors are prepared consistently to outspend America in space technology, then sharing present U.S. technology may merely accelerate their capacity to gain the lead.

However, some competition in space is likely to benefit all users of space services by stimulating innovation. One qualification of this statement occurs when economies of scale would permit savings through many nations combining for the use of a single facility, as, for example, was the case with Intelsat. Where such economies of scale exist, nationally based competition may lead to wasteful duplication, an event that may soon occur in the communications satellite field.[21] Such economies of scale arising from joint use of space facilities are particularly beneficial to smaller and less developed countries.

Another qualification may occur when some form of saturation exists, as is now occurring in the case of geostationary orbit/frequency slots. Where parking orbits are in short supply, unnecessary competition can be very destructive and can lead to accusations of discrimination and political conflict, as is also beginning to happen in the communications field. On the other hand, the shortage of parking orbits for telecommunications may be a temporary situation, which further advances in technology will alleviate. Although the United States has argued this in international forums, it has not backed the arguments with investments in the necessary technologies.[22]

In the long run, I think space operations should be placed in an international institutional framework as they move from R&D into routine services. A multinational corporate institution like Intelsat, even if it has to receive subsidies from participating governments, offers the greatest flexibility and incentives to meet the needs of users. There are growing political pressures against this approach, however. Many countries are planning for their own national communications satellites. In principle Intelsat participants must be able to convince the Intelsat board that their national systems will not injure Intelsat revenues. But the political pressures for the board to interpret this requirement liberally for the sake of peace may eventually become overwhelming.[23]

France is now expected to put up its own remote sensing satellite (*SPOT*) with capabilities exceeding in some respects those of the U.S. *Landsat D*. Services of this satellite will be available to all comers, with more flexible institutional arrangements and service assurances than the U.S. government seems likely to provide for *Landsat*. The European Space Agency (ESA) will offer an expendable launcher as an alternative to the shuttle, again with more flexible institutional arrangements. It already has several U.S. takers, although the recent failure of a launch of *Ariane* may change attitudes. At this stage these evidences of competition in space services may be healthy. They may stimulate less rigidity and more reliability on the part of the U.S. government. On the other hand, the prospects of competition in space, especially if based on heavily subsidized "national champion" institutions like Arianespace, could lead to a wasteful kind of competition in the future.[24]

In civilian space activities, it might be much better to have competition—where it is economically and technically desirable—take place between multinational entities, rather than on a national basis. Such entities could be owned and managed by groups of governments or involve combinations of private and public multinational ownership. Such an arrangement should lead to less destructive kinds of competition. A possible model exists in the international aircraft industry, where both development and production are increasingly becoming multinational, with joint ventures, co-production, and foreign sources of key components and technologies from many different countries. This is happening in the aircraft industry because both the technical and financial resources required to develop and deploy a modern commercial aircraft are taxing the means available to even the largest national economies.[25] This situation may be even more true of space ventures, increasingly as such ventures become more ambitious and the size of the U.S. economy shrinks in relation to the rest of the world.

Although it applies more evidently in the case of space applications, where an important fraction of total revenues may be recovered from users in several different countries, this situation could apply as well to science and exploration undertakings, where governments supply most of the resources. The cooperation between the United States and Europe (mainly Germany) on the *Spacelab* package for the shuttle has already set a precedent in this direction.[26] Furthermore, after more than a decade of false starts several European countries have finally achieved a viable and apparently successful mode of cooperating among themselves in the European Space Agency (ESA).[27] This cooperation could become a model for the full internationalization of space activities with an increasing number of adherents, much after the fashion of Intelsat. Because such international activities are more awkward to manage than national ones, in some small cases the "transaction costs" may not be worth the effort. On the other hand, reneging on priorities negotiated multilaterally between countries is much more difficult than on national budgetary commitments, and so the complexity of multinational operations may be largely offset by the greater commitment and stability that might be achieved in an international framework.

Space is inherently a global technology. The essential activity takes place in a global "commons" outside the sovereignty of any single nation; it requires a high degree of cooperation among installations all over the world, ranging from tracking stations to ground terminals and international data management centers. Thus, space is an ideal candidate for a multinational corporate form. In a sense the oil production and distribution system has become a global technology and has evolved such a multinational corporate form, but the parallel is far from perfect.

NOTES

1. Cf. E. M. Emme, ed., 1963, *The History of Rocket Technology, Technology and Culture* 4(4). (The papers published in this special issue of *Technology and Culture* were later expanded and republished as E. M. Emme, ed., 1964, *The History of Rocket Technology*, Wayne State University Press in cooperation with the Society for the History of Technology, Detroit, Mich. Subsequent references are to the original *Technology and Culture* issue.)
2. W. R. Dornberger, 1963, "The German V-2," ibid., pp. 393–409. E.g., "It would be foolish to think at that time the Germans had any definite idea about what would later evolve from their work. Yes, the initial, small group dreamed about long-range rockets and space ships. But they did not know and they did not care what would happen later. They just started with a power plant" (p. 396).
3. R. C. Hall, 1963, "Early U.S. Satellite Proposals," ibid., pp. 410–434. Project Rand, *Preliminary Design of an Experimental World-Circling Spaceship*, Santa Monica, May 1946, as follows: "To visualize the impact on the world, one can imagine the consternation and admiration that would be felt here if the U.S. were to discover suddenly, that some other nation had already put up a successful satellite." Also: "Since the United States is far ahead of any country in both airplanes and sea power, and since others are abreast of the United States in rocket applications, we can expect strong competition in the latter field as being the quickest shortcut for challenging this country's position."
4. Quoted in ibid., p. 419. *See also* D. Deudney, 1982, *Space: The High Frontier in Perspective*, Worldwatch Paper 50, Worldwatch Institute, Washington, D.C., August 1982.
5. J. P. Hagen, 1963, "The *Viking* and the *Vanguard*," pp. 435–451 in Emme, *History of Rocket Technology*. Cf. especially: "The letter from the Secretary of Defense [authorizing Project Vanguard Sept. 9, 1955] stated clearly that what was needed was *a* satellite [i.e. one] during the I.G.Y. which was to end in December 1958, and that the Vanguard program was in no way to interfere with the on-going military missile programs" (p. 439).
6. A. C. Clarke, 1945, "Extraterrestrial Relays: Can Rocket Stations Give Worldwide Radio Coverage?" *Wireless World* (October).
7. Quoted by Hall, in Emme, *History of Rocket Technology*, p. 434.
8. Cf., for example, J. Logsdon, 1970, *The Decision to Go to the Moon: Project Apollo and the National Interest*, University of Chicago Press, Chicago, Ill.
9. J. B. Wiesner, 1980, "Science and Technology: Government and Politics," in W. T. Golden, ed., *Science Advice to the President*, Pergamon Press, New York, as follows: "But when the President decided he had to have an aggressive lunar program as a political matter, I supported his decision and I didn't offer PSAC an opportunity to argue against it with the President" (p. 35).

10. D. R. Headrick, 1981, *The Tools of Empire*, Oxford University Press, Oxford.
11. R. Williamson, et al., 1982, *Civilian Space Policy and Applications*, Office of Technology Assessment OTA-STI-177, June 1982, Washington, D.C. Cf. especially chapter 4, "Development and Characteristics of the U.S. Space Program," pp. 81–104.
12. Public Law 85–568, *National Aeronautics and Space Act of 1958*, as Amended, 85th Congress, July 29, 1958.
13. H. Brooks, 1982, "Social and Technological Innovation," ch. 1 (pp. 1–30) in S. B. Lundstedt and E. W. Colglazier, Jr., eds., *Managing Innovation: The Social Dimensions of Creativity, Invention, and Technology*, Pergamon Press, New York. Cf. especially p. 3. For a major appraisal of the NASA management system, *see also* L. A. Sayles and M. K. Chandler, 1971, *Managing Large Systems*, Harper and Row, New York.
14. J. Adler et al., 1981, "In Space to Stay," *Newsweek*, April 27, pp. 23–36.
15. Williamson, *Civilian Space Policy*, p. 97.
16. Deudney, *Space: The High Frontier*, p. 13.
17. Ibid., p. 26. *See also* J. N. Wilford, 1980, "Riding High," *Wilson Quarterly*, Autumn 1980.
18. H. Brooks and J. J. Harford, rapporteurs, 1982, "Working Group Report on Communications," in J. Grey and L. Levy, eds., *Global Implications of Space Activities*, AIAA Aerospace Assessment Series, Vol. 9, American Institute of Aeronautics and Astronautics, New York, especially pp. 100–101.
19. W. H. Kincaid, 1981, "Over the Technological Horizon," *Daedalus*, Winter 1981, pp. 105–128, especially Table 1, p. 124.
20. Deudney, *Space: The High Frontier*, p. 18. *See also* T. Greenwood, 1973, "Reconnaissance and Arms Control," *Scientific American*, February 1973.
21. Brooks and Harford, "Report on Communications." *See also* J. L. McLucas, 1982, "Space Communications 1961–2001: Cooperation and Conflict," pp. 57–86 in same volume.
22. Deudney, *Space: The High Frontier*, p. 48, and Brooks and Harford, "Report on Communications."
23. McLucas, "Space Communications 1961–2001."
24. Williamson, *Civilian Space Policy*, pp. 183, 184.
25. J. Newhouse, 1982, *The Sporty Game: The High-Risk Competitive Business of Making and Selling Commercial Airplanes*, Alfred A. Knopf, New York.
26. Williamson, *Civilian Space Policy*, p. 180.
27. Ibid., pp. 176–177.

Comments

BERNARD A. SCHRIEVER
General, United States Air Force (Ret.)
Schriever & McKee
Washington, D.C.

Rather than try to respond to all of the points made in the Brooks paper, I will discuss the military involvement in space after World War II, because I was an active participant. In my comments I will touch on a number of things discussed by Brooks, including Apollo, management considerations, man-in-space, the civilian and military arrangement then and now, and at least some aspects of the future in space. I will cover three periods: the early *Sputnik* period, the more recent period, and the future.

Not very many people really know what considerations the military gave to space after the war. I was in the Pentagon in December 1945 after spending three and one-half years in the South Pacific. Prior to the war I had been in the research and development area as a test pilot and had gone to graduate school. After the war I went right back into the R&D business. General Hap Arnold, whom I had known for many years, was in charge of the Air Force.

In my opinion, we have never had a more visionary man than Hap Arnold. He understood the implications of the technological breakthroughs of World War II: the atom, rockets, electronics, jet propulsion, and so forth. Arnold said, at that time, that no future war will be fought like the last one. He understood the importance of the scientific community and what it had accomplished.

In 1944 Arnold asked Dr. von Kármán to be his personal science advisor, and he created a Science Advisory Board. He had von Kármán conduct a study called "Toward New Horizons," much of which was devoted to space. Arnold was also responsible for the creation of the Rand Corporation, which Harvey Brooks mentioned. Rand's first task was to determine the feasibility of a reconnaissance satellite. Another Arnold creation was a scientific liaison office to maintain relations with the scientific community. I had staff responsibility for the liaison office, along with the staff responsibility for Rand. From my perspective, serious thought was being given to space, at least at that level.

So very early on, we were thinking about space, not in a Buck Rogers sense, but in a sense of how it could be used to enhance national security. Even so, for the next decade—from 1946 through 1957—aside from the development of some rocket hardware, only staff study and analysis were done. We made little progress toward having a real capability in space. The launch of *Sputnik I*, of course, occurred in October 1957, and everything changed.

I had taken charge of the Air Force ICBM program in 1954. Early in February 1957 there was a symposium in San Diego at which I spoke on how the ICBM program had provided the resources and the know-how to launch payloads into space. I specifically outlined certain capabilities that we needed from a national security stand-point. I received a very negative reaction from Secretary of Defense Charles Wilson's office, and I was instructed not to use the word *space* in any future speeches.

I did have the responsibility for what space work we were doing at that time, and I was trying to get funding from the Pentagon. We had the "117–L program." It was really just a paper program, but it allowed us to identify satellite projects that would enhance national security. We had identified several potential programs, such as early-warning and reconnaissance. I managed to get 10-million dollars for space activities, but only for component development and testing and absolutely nothing for systems work. As a result, our situation was not conducive to moving rapidly into space in early 1957, al-though there was serious intent on the part of the Air Force to exploit space for national security purposes.

When *Sputnik* came along in October, the floodgates opened. Ei-senhower created the PSAC, with Dr. Killian as its first head. NASA came out of the 1958 Space Act. And the military created the Advanced Research Projects Agency (ARPA). During this period I shuttled back and forth from the West Coast to the East Coast, making presentations at the Pentagon or testifying before the Congress. And, I might add, I got myself in some hot water from time to time. Everything was happening very fast.

We were asked to accelerate the Minuteman program by one year, which, in fact, we did. We had an Initial Operational Capability (IOC) of the Minuteman within five years of the program's inception.

However, military space activity did not proceed smoothly after *Sputnik.* For example, in 1960, General White, the Air Force Chief of Staff, asked me to initiate a space study. I convinced Trevor Gardner to chair the study and assembled quite a group of outstanding scientists. It was completed in early 1961 and presented to the Department of Defense, but it was considered too provocative and was put on the shelf.

During the early '60s we had an antisatellite program, but it was canceled. We had started an early-warning satellite program called Midas, but that was also canceled. A communications satellite program called Advent was also canceled, and later on the Manned Orbiting Laboratory was canceled.

Certain strategic projects were carried forward, for which the Discoverer project was the forerunner. They were well-funded and highly successful, but emphasis was placed largely on NASA and its projects Mercury, Gemini, and then Apollo. Both Mercury and Gemini were dependent on military hardware. The Atlas and Titan missiles were used as boosters. The military and NASA worked together very closely during that period, and I think the special relationship we created still exists. Our programs have diverged, as Brooks has pointed out, but much of the technology is common, and we have benefited from each other's programs.

In the Apollo program, for example, NASA benefited not only from the standpoint of hardware development, but of management as well. An old friend of mine, George Mueller, who came from TRW to direct manned spaceflight, came to see me and said: "I'd like to have some of your people help us in the Apollo program." It may not be well-known, but about 25 Air Force officers, including General

Phillips, who became the program director of Apollo, were assigned to NASA. Indeed, we have been working with NASA for years. There are still a number of Air Force officers in NASA, and they are not working on Air Force projects; they are working on NASA projects.

Concerning the shuttle, I do not entirely agree with Brooks' comments. The plan, as I understand it, is that the Air Force will actually acquire, through NASA, shuttles that it will operate itself, so that the mixing of military and civilian payloads will probably disappear after a short period of time, I think within the next couple of years. I believe that mixing military and civilian payloads would create problems, but I know that the intent is to launch most military payload shuttles from Vandenberg Air Force Base. A facility is now being built there. The Air Force would take over operational control of those shuttles launched from Vandenberg, so that many of the serious problems cited by Brooks will not arise.

When I retired in 1966—almost a decade after *Sputnik*—I left the Air Force somewhat frustrated by the slow progress that had been made in applying space technology to the enhancement of the tactical capabilities of our military forces. I had in mind survivable, secure, and near real-time communications, command, and control of military forces—not offensive weapons, not even defensive weapons, but the enhancement of our ability to command, control, and communicate with our forces. Now, at long last, a very high priority is being given to the achievement of just this capability. I was on the president-elect's transition committee on science and technology. In our comments on space in our report to the President, we said: "A substantial space program is absolutely essential to national defense." We stressed communications, command, control, intelligence, and reconnaissance. We emphasized that space activity is also important to many civilian areas, and we suggested that, in those areas, the private sector should have greater opportunities and incentives to undertake an increasing share of the effort.

The recent policy statement by President Reagan also points out the importance of space to national security. I personally do not see any offensive weapons in space in the near future, not even in the next 25 years, but I do see space playing an increasingly important national security role. Since these assets are so vital to our military operations, space will no longer be a sanctuary. Anytime assets are as vital to military operations as our space systems are becoming, an enemy may attempt to destroy them. For that reason I believe that space will certainly not be devoid of military actions, that is, in the event of future military actions among major powers. And, as a military man who has been involved in several conflicts, I can assure you that the military is least desirous of fighting another war, particularly if it is likely to become a nuclear war. We are, in every sense of the word, trying to use space for peaceful purposes—in other words, to insure the peace. In that sense we are, in fact, right in line with the original objective of the Space Act of 1958, using space for peaceful purposes.

About man-in-space, the United States did not have men in space for quite a few years after Apollo, and the Soviets have had many, many more manhours in space than we. I believe we should not go overboard in spending large sums of money on man-in-space; the shuttle will insure manned experiments continuing through the next decade, which I feel will determine once and for all whether man has an important role to play in space. If I were asked today whether man has an important role to play, I would not be able to answer. My gut feeling is that in time he will, so we should continue to experiment and develop man's capability in space.

From a military standpoint, the 1980s will see space come of age in a tactical sense. That might not agree with Brooks' comments concerning the military, but I foresee real-time, secure, and survivable C^3I, that is, command, control, communications, and intelligence in support of military operations. Development of these capabilities has been given the highest priority; it will be a major challenge, and it will not be cheap. But, in my opinion, these things are absolutely essential. The shuttle and its follow-on activities will routinely put man in space so that, by the end of the decade, we should know whether man really has an important role to play there.

The third military possibility is in a controversial area: defense systems in space, such as particle beam and laser systems. No one country can be unique in this regard; therefore, we cannot avoid moving forward aggressively in the research and development of such systems. If we get beaten on this, it would have a very negative effect on our overall national security position.

Finally, I would like to comment on the commercial side, that is, the entry of private enterprise into the space field in areas where there has not been involvement so far—rockets, for example. The space shuttle is rather expensive, and it lacks certain flexibility. I am convinced that before long private enterprise will undertake the development of an advanced rocket. *Ariane* will be a tough competitor. We must develop advanced expendable launch-vehicle technology. I've been involved in certain discussions in this particular area myself.

There is a tremendous future for commercial applications in space, and there will be many payloads there. When I took over the ballistic missile program in 1954, I could not have anticipated that we would be where we are today in space, and that was 28 years ago. So I think that there will continue to be tremendous growth in space activities, particularly on the commercial side and in the vital role of enhancing our national security.

Comments

AMITAI ETZIONI
George Washington University

I have one principal comment and two minor ones. The first concerns a matter of some sociological interest. Harvey Brooks reached one major conclusion in his historical study of *Sputnik I*: that "the Soviet launching of an orbiting satellite in 1957, using a military booster, completely transformed the climate of opinion, not only in the United States but around the world. Not since the explosion of the atomic bomb on Hiroshima had a technological event had such an immediate and far-reaching political fallout." He does not complicate the situation by pointing to any evidence for that statement, nor does he give any examples of where that impact is to be found. [*Editor's note*: This passage was contained in the preliminary version of the Brooks paper, which was distributed prior to the October 14, 1982 symposium on which this volume is based. For the revised statement, see p. 8; see also the discussion following this paper.]

In fact, one of our leading political scientists, Gabriel Almond, in effect conducted a survey of what people thought about *Sputnik I*, before it was launched. In April 1956 he conducted a survey, which he repeated in November 1957, a few weeks after *Sputnik I* orbited. There is some sociological significance to the timimg, because most media events have a half-life of three weeks. That is, studies made at the apex—like the day the space shuttle *Columbia* landed in the desert—reveal a large public impact, but those made after two and one-half weeks reveal a quite different, and lasting, effect. Sociologists call this the "washout" effect.

Gabriel Almond asked people what they felt about the relative technological power of the United States and the Soviet Union. Those who thought the Soviet Union had moved ahead after the appearance of *Sputnik*—that is, the number of people who were favorably disposed toward the Soviet Union after that event—had increased by nine percent in Britain, remained the same in Italy, and declined by one percent in France and West Germany. These data, among others, suggest that the impact on public opinion, at least in those allied countries, was somewhat less than total.

In the other part of the world—the so-called underdeveloped countries—studies from Mexico City and Rio de Janeiro found that, in November 1957, 49 percent of the people in Rio de Janeiro and 33 percent of those living in Mexico City had not even heard about *Sputnik*. And that, of course, did not include the countryside. Moreover, in the United States, the *Milwaukee Sentinel* had a revealing headline: "Today We Make History." That was October 5, 1957; but the story relates to the Milwaukee Braves winning whatever they won. *Sputnik* was on page 3.

At the same time, Samuel Lubell, one of the most perceptive analysts of public opinion, conducted interviews all across the United States and found that Americans were repeating President Eisenhower's statement that *Sputnik* was a small grapefruit, which was not going to fall on their heads and was not much to worry about. This shows that, when there is a new event, previously unimprinted, the public tends to heed its leaders to a very large degree. Since initially the President chose, for various reasons, not to make much of it, it was not a big issue in the public mind—that is, until President Kennedy chose to make it an issue quite a bit later.

A more important point here is an issue that is still with us every day as we conduct our national affairs. What is the significance of world public opinion; what events impress it; and how can and should we react to it? The world society is not an American democracy in which the average citizen follows the news through open media and has an opportunity to act on his opinion in the political realm. In most parts of the world, as studies by Daniel Lerher and others show, the horizon of people extends only as far as their village. The national capital does not exist. World events have no significance. People live in the world of their village; they are preoccupied by the next meal, not world affairs. I did not put them there; I do not like it; but it is true nonetheless. To suggest that a technological jump, whatever its significance, will change their view of the jumping country simply does not square with what we know about human nature. Indeed, other studies (unpublished) done by the United States Information Agency suggest that the large investments in space by the United States and the Soviet Union run contrary to what these people feel national and international resources should support. Many of them would probably much prefer those resources were devoted to assistance to economic development in their countries.

My next two comments are much briefer. First, I do not believe it is completely accurate to suggest that there was "the greatest possible separation of NASA and Air Force" work from the beginning. I was happy to hear about Bernard Schriever's work with Jim Webb. The record is full of close and increasing cooperation between the Air Force and the space agency. Before the agency was formed, the Air Force had believed that there were some very important missions to be conducted in space, so it was not completely delighted when major resources were given to a civilian agency. Afterwards, the Air Force tried in a variety of ways to get its hands on the funding, or, failing that, on the missions, or at least on the outcomes, all of which led to a very large amount of mixing of missions.

For example, on January 22, 1963, the Defense Department, representing the Air Force, and NASA signed a formal agreement on Project Gemini, previously an exclusive civilian project. The agreement stated that the requirements "of both the military and the civilian manned space programs will from now on guide that project." The Air Force repeatedly used the NASA facilities, adopting the instruments for joint use. I see nothing nefarious in that, but it is not completely compatible with the statement that these were kept apart as much as possible, unless we interpret that phrase very loosely.

Finally, I would like to suggest one more item for consideration. I personally have argued over the last two years that there are many values not in outer space, but in near space. And I believe the program has been unbalanced from the beginning, because it has put its priority on deep space, where the elusive prestige value lies, to the neglect of near-space investment in communications, economic, scientific, and, if you wish, military systems. I believe the record should be examined and the merits of spending more of those billions in near-space be carefully considered.

Discussion

JOHN LOGSDON   The Milwaukee headline of October 5, 1957, triggered a memory of mine. At that time, while Bernard Schriever was worrying about how to get the Minuteman and Atlas and Titan to work, I was still in college in Chicago and, as college students are wont to do, looking for cheap beer. We figured that that night that Milwaukee won the World Series was the one night beer would flow in that city; so a group of us went 90 miles north in quest of cheap beer. But they raised the prices!

DICK PRESTON (Star Foundation)   Since most of you were in college when the *Sputnik* crisis happened, didn't you feel that it affected the younger generation, who actually affect the future, rather than the older generation? Do you think that the Russians were able to see farther into the future, and have affected more young minds around the world, by their technological advances?

HARVEY BROOKS   First, with reference to Amitai Etzioni's comments on the impact of *Sputnik*, I should have made it clear that I was talking about elite opinion. There is very ample documentation—interviews with political leaders, newspaper stories, and so on, contemporaneous with the launch of *Sputnik*—which indicates that there was a profound shock to elite opinion around the world. I did not mean general public opinion as revealed in public surveys. Etzioni's point is perfectly well-taken, but it is irrelevant to what I had to say, although I did not say it clearly.

Sputnik did have a very large impact on the imagination of young people, especially those who were destined to become members of what Jon Miller [Northern Illinois University] and others call the

37

"attentive publics" for science and technology. We have to look at the whole question of the climate of opinion in terms of the attentive publics. It is they who tend to have a great influence on political events.

AMITAI ETZIONI   Harvey Brooks is correct. A study of interviews conducted that particular week with world leaders would reveal that they did express dismay and shock. But if we had followed them two and one-half weeks later, to see if these world leaders had left NATO, voted against us in the United Nations, or sent one fewer soldier to Korea, there would have been no discernible effect. It did not become a public opinion issue until May 1961. Between 1957 and 1961 there was no political fallout.

DICK PRESTON   The world is controlled by the one percent of the population who are the most forward thinking or have the greatest intelligence of forethought. There will always be detractors who will say that the world is flat or that new technology really is counter to man's best nature, but man is a technological beast. The very fact that he picked up the first stick changed not his hand, but his brain, and a nation that forgoes a rigorous prosecution of technology is a nation that becomes a backwater—an interesting curiosity of history. I would like to ask the military, civilian, and anti-technology people, why we have allowed our country to become so backward.

AMITAI ETZIONI   I am not anti-technology; I do not believe the world is a village; and I do not believe man is a technological beast. To suggest that technological resources might be allocated differently from investing a hyperportion in the Apollo project is not to be any of those things. We can all be in favor of technology, even the U.S. Air Force, and still not put 25-billion dollars into Project Apollo.

NOEL HINNERS   On the role of man in space, I have frequently claimed that some 90 percent of the beneficial aspects of space comes from 10 percent of the expenditures. The other 90 percent goes for the manned activity. However, having spent time at the National Air and Space Museum, I have seen another aspect, which I think we have all missed. It has nothing to do with a utilitarian role for man in space. Those interested in the space program who come to the museum have a view of humans in space that is very simply an extension of themselves into that environment. Neither they, nor I, are very good at describing this, but it is a human endeavor, which is perceived as such; it has nothing to do with the practicalities of the matter.

BERNARD SCHRIEVER I do not quite understand the 90 percent versus the 10 percent because the Air Force has never had a man-in-space program. All of our programs, both the highly classified and the others, have been for unmanned satellites. The Defense Department will be spending more on space activities this coming fiscal year than will NASA, so I think your numbers are wrong.

NOEL HINNERS I am saying that most of the benefits, 90 percent of the benefits, come from 10 percent of the expenditures, that is, from the automated programs.

JOHN LOGSDON But the point is if you put the military together with the civilian budget, the automated programs are a lot larger than 10 percent.

BERNARD SCHRIEVER As a matter of fact, the Air Force, which has been the major military service in space, has not yet come up with what I would consider a legitimate basis for putting man in space for military or national security purposes. However, my gut feeling is that man is going to play a very important role in space at some point; I just do not know when that will be.

JOHN LOGSDON The lack of a rationale is not from the want of trying.

BERNARD SCHRIEVER That's right. We've tried.

GEORGE FIELD (Harvard-Smithsonian Center for Astrophysics) You said, General Schriever, that you did not see, for the next 25 years, a requirement for offensive weaponry in space. Yet you mentioned command and control systems in space and, along with those, the necessity for deploying laser and particle beam weapons in space. If, in fact, these C³I [command, control, communications, and intelligence] facilities are deployed, and if the Air Force—and presumably the Soviet Union—believe that it is also necessary then to deploy beam weapons in order to counter these facilities, how can we avoid an offensive confrontation, which would presumably escalate very rapidly into a nuclear war?

BERNARD SCHRIEVER   I did not say that we should deploy laser weapons. I said laser or particle beam weapons would be a major breakthrough and could change the whole force structure of the military; therefore, we have to conduct research and development on them. Right now, I do not see an early solution to the R&D problems associated with such weapons. But we have to put high priority on the R&D, because development of such weapons would be a very dramatic breakthrough, and it would be very much to our disadvantage if the Soviets developed them first. C³I systems are not offensive systems, and we are doing other things to develop an antisatellite capability that is protective, not offensive, in nature. Antisatellite devices would protect our satellites; they would not be offensive, in the sense of attacking things on earth.

GEORGE FIELD   Then I request that we focus on the C³I capability. Indeed, it seems that it would be a great advantage to either power to knock out the capabilities of the other; therefore, that calls into being the requirement for further offensive capabilities. Do you see how those offensive capabilities could be deployed and possibly even used without turning the situation into a nuclear war? If the C³I systems are, in fact, knocked out, it would thoroughly compromise the ability to wage nuclear war.

BERNARD SCHRIEVER   I think you are saying that any kind of war between the major powers is going to be a nuclear war. I do not agree with that.

GEORGE FIELD   But do you agree that if the C³I were knocked out or even threatened, there would be an enormous incentive for launching a first strike?

BERNARD SCHRIEVER Military forces must have command and control; they are essential. All other forms of command and control are vulnerable; they are going to be targets in the event of a major war. I think you are saying that *we* should not develop them because if they are attacked that will insure an all-out nuclear war.

GEORGE FIELD I was not suggesting any logical conclusion from this. I simply wanted to understand your comment, which seemed to suggest that we can proceed in this area without getting into offensive systems. I do not think that is possible.

BERNARD SCHRIEVER We will get into defensive systems, but not offensive systems. I do not see involvement in offensive systems that are nuclear in nature, which would be launched from space. I would call them strategic offensive systems.

JOHN LOGSDON What if Secretary of Defense Charles Wilson had not been so strongly anti-space in 1957 and had not forbade the von Braun team in Alabama from launching something into orbit in midsummer of that year? What if Von Braun's technical caution had not required an extra flight with a chimpanzee in March of 1961 (I think), which should have been the flight of Alan Shepard, and Shepard became the first man in space, rather than Yuri Gagarin? What if this country had not been responding to a Soviet challenge? What kind of space program might have evolved? In other words, what are the merits of space development on its own without the broader competitive challenge that space began to symbolize over the last 25 years?

HARVEY BROOKS There would have been considerably less pressure for a manned space program. We would have been much more likely to have had a plan for an automated landing on the moon, rather than a manned one.

JOHN LOGSDON   What about the military programs; would they have looked much different?

BERNARD SCHRIEVER   We already had the capability to move forward into space. We had the studies, the analysis, the technology, the resources, and the know-how to put satellites into low earth orbit. We could have taken the lead much sooner in those programs and provided ourselves additional military capabilities. Without *Sputnik*, we probably would not have launched a major manned program, had we moved forward first.

AMITAI ETZIONI   One consequence would have been to the exhibits at the National Air and Space Museum. But the serious point that needs to be made is that, when another country challenges, you do not have to respond on the same front unless you choose to. That allows the other side to define the agenda, which is half the battle. What if we had increased the Peace Corps by a factor of 10 as a response?

JOHN LOGSDON   President Kennedy did go through that kind of exercise, thinking, for example, about nation building in Kenya, desalinization, or some other large technological demonstration other than space. But he decided that the terms had been defined so closely that he would have to respond in space.

KERRY JOELS (National Air and Space Museum)   Would General Schriever comment on our apparent philosophy of using fewer and more complex systems in space than the Soviets, who have cheaper, more easily replicable systems?

BERNARD SCHRIEVER   In the military sense, we have gone to highly complex, technically sophisticated satellites, largely because the programs aiming at enhancing our military capability and our national security have been strategically oriented. Very little consideration has been given to how they might be used in wartime or in a tactical sense. As we move forward and recognize the necessity of having both secure communications and survivability, we may go to different types of satellites. Of course, I am referring to military operations, not civilian communications satellites. It will be very difficult to have complete survivability of our space assets, and that whole matter is under very serious study at present. A number of different ways are possible. One might well be more satellites, simplified for a special task.

KERRY JOELS Concerning Etzioni's comments, satellite communications can now reach into the village with information, thus bringing the village into the political arena for a government. It provides a mechanism for capturing the hearts and minds of its remote minorities or majorities. What do you see happening as a result of this communications capability?

AMITAI ETZIONI I can easily accept that, in the not too distant future, we will develop the technology by which a satellite will be able directly to address a cheap local TV instrument in the home without having to go through relay stations, and that, in effect, we will be able to address the world more or less at once. But there are not many things that a scientist can say about the likely effect of that technology.

Exposure of people to a message à la Madison Avenue is highly effective only if they have a favorable predisposition to the message in the first place. That means, if people want to buy cigarettes and you spend 100-million dollars telling them to shift from one cigarette to another, as long as there is no difference between the two cigarettes, you can shift them. If you try to go a little beyond that, like moving them from regular coffee to instant coffee, you already have a crisis. And if you go anywhere beyond that, like trying to change their minds on their views of the Soviet Union or the United States, forget it. We have very good data to show that. So, the idea that you can advertise emotional or ideological messages to people and turn them on is not supported by evidence. The Soviet Union has had complete control of its educational and communications media for 50 years, and it has not brainwashed its people. Current events in Poland are additional evidence.

In the end, what people care about is their values; they are just like us. Suppose you were back in one of those villages and two messages appeared on your TV: one announces that the United States has just put a whole football team on the moon, and the other says that they shipped a bicycle for every villager down the road. Which message will make you jump higher?

MORRIS FRIEDMAN (Library of Congress)   How is it that the Russians have had a very extensive and successful space program for the last 25 years? They must have spent a great deal of money to launch their 1412 satellites and, with *Salyut VI* and *VII*, to keep a man in space for so long. Why have we not responded to this challenge? If we are going to have a man in space, for either defensive or offensive purposes, we will have to do a lot of research of one sort or another to keep him up there, to enable him to operate up there efficiently, just as the Russians are now doing with the *Salyut* program.

BERNARD SCHRIEVER   There is a debate going on now within the Administration on creating a manned space station, as perhaps one major NASA program of the future. Obviously, this is a controversial issue.

As far as the Air Force is concerned, there is a limit on the amount of funds that are available, and we are more or less dependent on NASA to do any manned space projects. We are becoming directly involved in the shuttle, and we are very happy to have it. I hope that we learn a lot of lessons with the shuttle over the next 10 years.

SECTION 2

The Practical Dimensions of Space

Introduction

WALTER SULLIVAN
The New York Times

At the National Academy of Sciences just 25 years ago, a conference was taking place that ended rather dramatically with the launch of *Sputnik I*. Anatoli A. Blagonravov and his colleagues, who were participants in that conference on rockets and satellites for the International Geophysical Year, had been dropping hints right and left that something was going to happen. It was a rather sad day for me as a journalist because I had been hesitant to report such imminence. On the day I finally wrote it, the story was never printed. That was the day it became a reality.

We should not have been quite as surprised as we were. We were in the habit of looking down on the Russians. In those days visitors to the Soviet Union had difficulty imagining how they ever got anything off the ground, particularly if the visitors had had anything to do with Intourist [the state travel agency].

But there was Tsiolkovsky who, at the turn of the century, dreamed of going to the moon. There was the formation of GIRD, an acronym for the Soviet Union's rocket society back in 1929. The Russians were quite advanced in their rocketry during World War II. They did not develop V-2's, but they did have the Katyusha rocket batteries that were the nemesis of the German forces at Stalingrad. And before the launch of *Sputnik I* there were plenty of hints. The Russians had announced the *Sputnik* tracking system so that people in various parts of the USSR, which meant anybody anywhere else who read about it, could track the spacecraft as it went by.

I was amused by John Logsdon's mention of that early White

47

House paper, which took pains to explain why *Sputnik* stayed up. It reminded me that perhaps the first person in history to discuss the launch of an artificial satellite was Sir Isaac Newton. An illustration in his *Principia* (second edition), published in 1713, showed how to launch an earth satellite. It portrayed simply a cannon on top of a very high mountain. Of course I do not think Newton went into the problem of atmospheric drag, whether the mountain would be high enough for the cannonball to be above the atmosphere. But it was a beautiful illustration of the orbital phenomenon. He showed the cannon firing a succession of cannonballs with increasing velocity. Each went out a little farther around the curvature of the earth, until finally the trajectory was sufficient to keep it circling the earth forever.

The American space program, especially the manned lunar landings, was also anticipated as early as 1865 by Jules Verne. He was remarkably perceptive. In his introduction, Verne said that if anybody was going to the moon, it would have to be the Americans. The Italians were the musicians of the world; the Germans were philosophers; but the genius of mechanics rested in the United States. Verne said the lunar voyage would be financed in considerable measure by the Russians. In fact, they put up 368,733 rubles. Jules Verne was remarkably well informed. He also explained what happened on this mythical trip when the dog the travelers had taken with them died. They decided to throw the animal overboard, but the dog stayed with them because it had the same momentum as the vehicle. For the rest of the trip, they were stuck with a dead dog floating just outside the window.

As far as the subject of the second symposium section is concerned, "the practical applications of space" were also anticipated—this time by Arthur C. Clarke in his 1945 paper on communications satellites. For some reason Clarke thought such satellites had to be manned. He envisioned a triad of satellites, equally spaced around the equator, so that they would have complete coverage of the earth's surface and would form a network for relaying communications. It was many years later before we really began thinking seriously about such practical uses for spacecraft.

There was discussion earlier of the impact of *Sputnik*, of the Apollo project, and of the space program as a whole on the public. Certainly, in terms of personal wonder and amazement, the sight of the first *Sputnik* and of the satellites that followed left an indelible impression. Any of us who went out at night to see them go by will probably never forget the experience. It was an awesome reality, not something you read in the newspaper. You went out there and looked at it. In fact, in *The New York Times*, for a long time, we published everyday the schedule of when *Sputnik I* would go by. We believed it was something thousands or millions of people would like to see.

Concerning the Apollo landing on the moon, I remember the story, and I believe it is true, that when Neil Armstrong was about to descend from the lunar module, and the television camera was set up and the feed from the Apollo center to all the television networks was all arranged, the authorities in Moscow decided not to show it on Soviet television. But the members of the Soviet Academy of Sciences, very influential ones politically, went to the Politburo and objected, saying it had to be shown. So it was. And in places like Beirut, where I had friends, and elsewhere in the world, people telephoned Americans to congratulate them on that achievement. It was a great event at the time, so we should not denigrate it now, although it has been forgotten to some extent.

There are three contributors to this section. The first, Simon Ramo, needs no introduction. He was cofounder of TRW, which of course stands for Thompson, Ramo, Wooldridge. He was chairman of the Committee on Science and Technology under President Ford, and has been on so many advisory panels for the present government as well as for previous ones that it would leave no time for anything else if I listed them all. Ramo was born in Salt Lake City, Utah, in 1913, attended the University of Utah and then Caltech, where he took his Ph.D. in electrical engineering and physics; he then worked as a research engineer at General Electric Corporation from 1936 to 1946. He was vice president and director of operations for Hughes Aircraft for the next seven years, and then served as executive vice president of the Ramo-Wooldridge Company from 1953 to 1958. He was scientific director of the U.S. Intercontinental Guided Missile program from 1954 to 1958—the critical years under scrutiny at the moment—and then became head of TRW. Simon Ramo has continued to be the "grand old man" of the space business ever since.

Roger Chevalier is executive president of Aerospatiale, a large French company engaged in an amazing number of enterprises, including producing the Airbus, helicopters (not only in France, but in Texas), the now-famous Exocet missile, and the *Ariane* rocket that has begun to compete with the space shuttle as a launcher of telecommunications and other satellites. Because of this mix of experiences, he is exceptionally well-qualified to comment on Simon Ramo's paper. Chevalier was born in Marseille in 1922 and was educated at such institutions as the École Nationale Supérieure de l'Aeronautique. He was recently named president of the International Astronautical Federation (IAF).

Finally, Edwin Mansfield looks at things from a different point of view. He is an economist, who is a leading authority on monetary and fiscal theory. His recent article in *Science* magazine presented an extensive analysis of tax policy and how it can stimulate (or not stimulate) innovation. Perhaps this cost-benefit analysis can be applied to some of the problems in the space program. Mansfield is professor of Economics at the Wharton School of the University of Pennsylvania. He was born in Kingston, New York, in 1930, took his undergraduate degree at Dartmouth, his graduate degrees at Duke, studied statistics at the University of London, and was a Fulbright scholar in Britain. From 1955 to 1960 he was at the Carnegie Institute of Technology; then he served at Yale, Harvard, Caltech, and the Institute for Defense Analysis. He has been chairman of the USA-USSR working group on the economics of science and technology.

The Practical Dimensions Of Space

Simon Ramo
TRW, Inc.

Space, its exploration and utilization—to use a peculiarly apt figure of speech—is out of this world. A quarter-century ago space burst forth as an arena for a "Science Olympics" between the United States and the Soviet Union. Now it is a region for potential warfare and an indispensable tool for disarmament verification and the prevention of war. Space is a multibillion-dollar civilian growth industry and an inexhaustible frontier for scientific research. How did it happen that the U.S. space program, which started so suddenly, immediately rated so high a priority as to acquire an annual budget in the billions, a brand new agency reporting to the President to manage it, committees devoted to it in both houses of Congress, and more attention by the world's communications media than any other science or technology program in history?

THE SPACE RACE

It is no mystery. The launching of *Sputnik I* by the USSR in October 1957 surprised the world, but it shocked the United States. We knew the Russians excelled in ballet and caviar, but when the proper time came to launch an artificial moon, we Americans expected to be the ones to do it. We were already preparing a modest instrument package to be sent into orbit as part of the International Geophysical Year. So when the Soviet Union upstaged us, we were insulted and alarmed. If they could do this, we could expect them to abandon the inferior status we had conveniently assigned them in science and technology in general, and they might outdo us in military weapons systems as well.

We reacted emotionally to the fear and the dare and the Space Race began, which led to our inventing the major event in the new Olympics: a manned lunar landing. This world spectacle, the boldest space feat then imaginable, became symbol and substance for our regaining the lead, not only in space, but in all science and technology. After the Soviet sputnik blow, interest in education in technical fields ballooned, and the government increased all its R&D budgets. Thus, not only was space technology launched as a new priority category, it became the spearhead for accelerated efforts on every science and technology front.

It was not that, having looked carefully at all that science and technology might make possible, and noting our most urgent needs and exciting opportunities, we decided the moment had come to explore space avidly, all rocket nozzles aglow. Nor did we opt for space in the late fifties merely because our advances in technology finally enabled us to do so. Once the grand-scale program to provide an ICBM had matured, we automatically had the ability to orbit a substantial payload. Our ICBM program, begun some four years before the launch of *Sputnik I*, had developed the entire range of technology: large rockets and matching fuels, light yet strong structures, electronics for control and guidance, reentry techniques, production lines turning out reliable quantities of hardware components and assemblies, test instrumentation, and large-scale launching and tracking facilities stretching out over the Atlantic from Florida. Because the earth is round, an ICBM capable of delivering a substantial weight accurately on a target half the earth's circumference away could easily be put on a trajectory to cause a somewhat smaller payload deliberately to overshoot, miss the earth entirely, and go into orbit. The ICBM technology was extendable to place a human passenger in orbit and provide him with oxygen, food, reasonable comforts, suitable communications channels, and protection during reentry.

Thus, by the late 1950s, we could commence satellite and other manned and unmanned spacecraft projects. However, if the Russians had delayed *Sputnik* for years, our space program would have started later at a less frenzied pace. The first major goal, for instance, would have been to meet a military requirement with a spy satellite, or to develop commercial projects with immediate return on investment, such as intercontinental television and telephonic links by satellite. Scientific curiosity about how biological matter—a man, for example—might react to a gravity-free environment would have stood in an orderly line along with inquisitiveness at other frontiers of knowledge.

Indeed, those who argued that the enormous funds for the manned lunar landing project (nearly 100–billion 1982 dollars) should be spent on other things, such as broadening astrophysics research by observations from unmanned spacecraft, advancing microbiology or high-energy particle physics, or seeking a cancer cure, underestimated the backing for a direct contest with the Soviet Union. If we had wanted only to examine the moon scientifically, we surely could have done so more quickly and cheaply by sending instrument packages there, even including a device to pick up moon rocks and return them to earth automatically. But putting humans into a spacecraft and sending them to the moon satisfied our psychological need as no other competitive project could. The successful manned lunar landing replaced the American public's feelings of newly found inferiority with newly confirmed superiority. Concern turned into exhilaration. As a momentous achievement, visiting the moon pushed all else temporarily into the background. American astronauts placed there seemed to reestablish all Americans as leaders and pioneers.

MILITARY SPACE

After the sputnik surprise, the American public quickly associated the technological prowess demonstrated by the Soviet Union with a military threat. Some warned that the moon must be captured immediately and turned into a solely American platform from which to bomb the earth, or to be used as an invulnerable hiding place for nuclear weapons. They warned that the moon is the high ground, so whoever controls the moon controls the earth. If we do not move fast, what will we find when we land on the moon? Russians.

Twenty-five years later, the moon still has no war role. Space has become essential for certain military functions, but the moon—which is the low ground to one on it looking up at the earth—is not a sensible military base. Its role ranges from hopelessly uneconomic to irrelevant when compared with that of custom-designed artificial satellites. Its orbit relative to the earth is unsuitable. Its back side as a place for storing missiles and nuclear weapons, which might be expected to survive any attack and deliver massive retaliation, is not necessary and not even advantageous. (If we insist that weapons storage locations be extremely expensive to reach and inhospitable to humans, as though this equates to invulnerability, such places already abound on earth at the North and South Poles and the vast ocean bottoms.)

In today's military operations, reliable communication between points on land thousands of miles apart is a fundamental requirement, as are message transfers among ships, aircraft, and land stations. Satellites sometimes offer the only route for military communications signals. They can expose to observation the entire land and ocean surfaces on earth, as well as the atmosphere and space. Instrumented packages in space can photograph the earth, probe it to learn what is there, detect enemy communications and all manner of radiation and fallout, and provide tracks of missiles and spacecraft. Measurements from space can disclose tests that might otherwise go unnoticed and enable one nation to gain a great military advantage over another. Sensing from space enables a country to distinguish peaceful excursions into space from those that cannot have any but a military purpose and might expose an attack as occurring or imminent. Moreover, space can be used for countermeasures to interfere with an attack if it is launched.

Reducing the danger of nuclear war by arms reduction pacts is urgent, but such reduction is not credible without a system in place to provide continuous information about the war-related activities of the world's nations obtained through indirect (not on-site) monitoring. Verification of adherence to agreements has to include satellite-based sensing. Of course, space systems for superior navigation of ships and airplanes, and a similar enhancing of our ability to observe and predict the weather, are at least as valuable in military as in civilian operations.

These uses of equipment in space are not limited to strategic warfare applications. In the NATO area, over a shorter but still significant geographic span, the multiple functions of communications, command, control, reconnaissance, intelligence, detection, and warning are mandatory. Sole reliance on information relayed directly between ground and/or aircraft stations is insufficient. Satellites in geostationary orbit (the satellite's rotation about the earth synchronized with the earth's rotation so that the satellite appears stationary) provide better means to handle many of these tasks.

Since placing appropriate equipment in space offers military advantages, necessarily the military will consider removing the enemy's equipment from space in the event of war. In peacetime, space will be populated by apparatus placed there by many nations. If world war should come, hostilities would probably spread quickly to space. Some actions there might even precede land engagements, since eliminating warning and communications capabilities might be essential to the initiator of hostilities.

Whatever a spacecraft is doing for the military, it should have coding and anti-jam features. A military communications satellite thus typically uses circuit techniques beyond what is needed for a civilian communications satellite, such as for television or telephone transmission. The payload sent into space by the military ideally should be of extremely high quality, optimized as to function, very reliable, survivable under adverse conditions, and possessed of long life. The cost per pound to orbit a payload is high, so microminiaturization is especially important. Withstanding the launch vibrations and acceleration and enduring the space environment add unusual requirements for ruggedness and thermal and radiation immunity or shielding. Space-based military systems accordingly are near the limits of the scientifically and technologically possible.

There have been some proposals for civilian space projects that exceed existing military projects in boldness, performance, scope, technological reach, cost, and complexity. However, military implementations will often precede the use of similar techniques in commercial areas. Military space projects thus serve both to provide for defense-related applications and to push forward the technological frontiers for eventual civilian space applications.

COMMERCIALIZING SPACE

With today's technological capabilities, we can place equipment in orbit for substantial periods to enhance the operations of our surface civilization. Examples are:

• telephony between continents, and over large earth spans on the same continent, through satellites, as a link often more economical and higher in capacity than cables or other alternatives;
• television relaying, by satellites, point to point over all the earth;
• airline navigation and traffic control, satellites acting as signaling, artificial stars moving in precise and predictable orbits, in communication with ground computers and airborne transmitter-receivers, for higher traffic capacity and more accurate and economic locating of aircraft in flight;
• navigation and communication for ships at sea;
• weather information and prediction by satellites monitoring the dynamic characteristics of large land, atmospheric, oceanic, and space regions and reporting the data instantly to ground data-processing stations;
• earth resources satellites to scan for mineral, forestry, water, and agricultural resources, fishing, and pollution information to improve discoveries, warnings, and utilization;
• computer-to-computer information transmission between ground points to maintain the logistic, scheduling, accounting, and control data of industry and to provide professionals with access to stored information; and
• direct-to-rooftop broadcasts, together with local cable systems, to bring wideband, multiple-channel programs to mass audiences.

In another decade, hundreds of millions of individuals in the noncommunist world will probably average more than an hour a day in some activity involving a satellite—talking on the phone, watching television, traveling in an airplane, acquiring data at the office, being educated. Putting a modest value of only a few dollars per hour per person on this service leads to an estimate of more than 100-billion dollars for the annual revenues. This range of revenue suggests an investment to make it possible in the 100-billion-dollar range, and, if the installations are economically sound, yearly returns on the investment of around 10-billion dollars. This is greater than the average annual expenditures made in space by the government in the past, which means that we may be nearing a period of net positive financial benefit from the nation's investment in space through civilian commercial activities alone, even without adding any national security contributions.

But the real impact of these satellite applications may go far beyond mere financial investment returns. Consider the impact of conventional telephony in the last century. Without telephones, communication—in the sense in which Americans, especially in business, have organized around it—would not have been just more expensive, it would have been impossible. The ultimate effect of satellites in opening up new dimensions of communication will almost certainly be just as revolutionary, transcending purely economic measures.

Even for these commercial applications, the potential market is far from filled. More than a billion-dollars' worth of commercially owned satellites are in orbit today. Before the end of the 1980s, a dozen or more U.S. communications satellites will be authorized yearly by the Federal Communications Commission (FCC), and the rest of the noncommunist world will require a like amount. The available parking orbits and radio bands are already crowded, and future assignments will involve considerable intergovernmental negotiation.

There will be other new commercial applications as well. A number of ambitious proposals already have acquired enthusiastic promoters. Even if they never reach fruition because of inadequate economic or other rationale to justify their backing, they are worth mentioning to show the scope of interest and imagination at work. For example, one proposal involves creating huge structures in space to capture massive amounts of solar radiation for conversion to microwave power, which would be transmitted to earth by radio beam. Here the received power would be converted to 60-cycle electricity for use in the nation's electric power distribution network. Even the most practical designs to implement this idea still are so bold as to astound those with experience in engineering and financial investments. The National Research Council has said that a satellite-based solar power system, one providing a substantial fraction of our total electric power demand, would be the most costly and complex undertaking, civilian or military, ever attempted, with total costs amounting to three-trillion 1980 dollars. The structure would encompass about 25 square miles, and 20 years would be required before a useful demonstration could be made. A large-payload repetitive space booster would bring the parts and astronaut-assemblers to space in countless roundtrips. The unprecedented amount of energy and resources needed for the construction, the risks, the potentially severe environmental impact, and the numerous technical problems have caused most who have examined the concept to give it low ratings when compared with other alternatives for meeting future energy needs. The government has so far commissioned only preliminary studies.

Another potential space activity, also speculative but not as enormous and far less risky, involves manufacturing in space. Certain classes of materials—unusual semiconducting crystals, superior pharmaceuticals, revolutionary chemical catalysts, exceptionally high-purity glass, greatly more precise ball bearings—cannot in theory be formed while gravitational forces act on the process. In principle, the fabrication could be conducted in the weightless environment of space. This idea obviously would apply only to materials whose unusual characteristics equate to unusually high market values. Manufacturing in space probably would require a large manned facility there. In fact, a substantial manned space laboratory may have to be in operation for years before the potential of manufacturing there, automatic and unmanned or man-operated, can be evaluated. At this time only a preliminary, experimental program is underway.

Another speculative application is the mammoth communications satellite. With high transmission power and the potency of its large antennas for picking up weak ground signals, very small transmitter-receivers and antennas on the earth's surface would suffice. Such a system could literally put every human in touch with every other one if each had only a wrist radio, an antenna the size of a soup dish, and a direct line-of-sight path to a satellite. Such a system is technologically possible, but the present need does not justify the cost, so the development is for the future, if ever.

Super-large satellites have other possible applications. For example, they might carry equipment that could detect tiny amounts of nuclear radiation. This would help pinpoint the location, and police the movement, of radioactive materials. The satellites also could pick up warning and distress signals of many kinds and retransmit them to speed help.

The economic benefits of such commercial applications of space result from putting equipment into near-in space. What about space farther out? What about the moon? So far, our explorations have eliminated the moon as a cheap source for commercial-grade green cheese and disclosed little to suggest economic bonanzas from the moon's new accessibility.

What about the more distant planets? Nothing we know about them now would suggest any practical venture for commercial exploitation. What about other stars, perhaps other galaxies? The techniques are not even remotely apparent for reaching them. Those means are more likely to be found, if at all, by indirection—discoveries made while pursuing unrelated research—than by a brute-force effort to keep going up and away with the hope that the means to reach distant stars will thereby be revealed. Efforts to improve the breed of racehorses did not lead to the invention of the automobile.

But the earth often appears to the average human being on it as an exceedingly crowded place. The world's population probably will continue to grow, but the planet's surface is of finite size. We have built skyscrapers in an attempt to push into a third dimension, but then the two-dimensional traffic problem in getting from home to work and back becomes worse. Now, in one field of human endeavor—space conquest—we can speculate about an increasingly accessible, infinite three-dimensional volume.

The resources here on earth are finite. Even if we use them efficiently, a continuing population increase at a rate greater than our acquisition of resources may limit life on earth by the need for ever-decreasing allotments per individual. Space offers new sources of energy and the potential of resource supplies from interplanetary space and many planets and moons. Space is a new frontier beyond which is much more than a new continent. We cannot even guess at the scope of it, and it may eventually lead to unforeseen commercialization.

HUMANS IN SPACE

When America launched its space program, it was an emotional reaction to the sputnik blitz. Not surprisingly then, humans in space dominated the first round of the competition. The space age had to be symbolized by humans heading the conquest, there in person, participating directly.

Sputnik happened at the same time as a related battle over the role of humans. The arena was the atmosphere, and the issue was the need for a man in an aircraft. At World War II's end, the most glamorous military man was in the cockpit—a fighter or bomber pilot. But soon guided missiles entered and usurped a part of such important duties as shooting down enemy craft, delivering tactical and strategic bombs, and attacking enemy land and sea forces. For many key military missions, missiles are faster, more effective and economical, safer for military personnel, and can operate under a broader range of environmental conditions.

The first phase of America's entry into space involved man very conspicuously, but it failed to establish a substantive continuing role there for human beings. If we want readings of physical phenomena in space, instruments can do a better job than humans. Automatic communications equipment will bring the information down to earth more accurately, completely, and quickly. If we want to know how the earth appears from space, or what the moon looks like to an observer there, man-made devices can pick up anything human eyes can, and in greater and more focused detail. The same is true if our purpose is military intelligence and reconnaissance. Finally, keeping any spacecraft on the proper trajectory and in stable flight is a function best suited to instruments.

A human being in a spacecraft drastically complicates the project. Providing for the safety, health, and comfort of an astronaut from takeoff to a return landing narrows the range of permissible risk-taking and adds weight, cost, and time to the exercise.

A competent systems designer will seek the right combination of humans and machines to accomplish any task sensibly. To rely totally on man's hands and backs, or brains and senses, with no associated mechanical or electronic tools is usually an extreme, not the optimum. Conversely, completely automating everything is rarely best. *Homo sapiens* is produced by cheap labor and, given its combination of sensing, intelligence, and motions, has a low annual maintenance cost. But mechanical devices can exert tremendously more force and withstand a much more severe environment. A human being can multiply a one-digit number by another at the rate of only one per second. A computer can multiply two multiple-digit numbers in less than a millionth of a second. People and machines each have their places, in space as on the ground.

For civilian and military applications in space, few essential functions have emerged so far that absolutely require a person's presence. The Apollo flights, which featured astronauts, came to a dead end when enough successful landings had taken place and the program had accomplished its psychological mission. Public interest waned and NASA budgets drifted downwards. Civilian space applications, without humans, received all the attention of the private sector, and unmanned satellites commanded first priority for the military. Pure research in outer space continued, but attention turned to the more interesting exploration of the other planets with unmanned, instrumented spacecraft.

A package was landed on Mars which, without the aid of a human, scooped up Martian surface matter. Then highly automated, microminiaturized laboratory and computer-communications equipment within the package processed the material, examined it carefully for signs of life, and transmitted the results back to earth. The rapid advance of information technology increasingly is making possible more sophisticated automation of information handling—whether it be sensing, processing, on-board control and navigation, or communication back to earth—at less cost, with less weight, and with increasingly high reliability. This militates against the employment of humans in space.

Until the program to develop the shuttle began, America had accepted a seven-year hiatus in placing humans in space. The shuttle arose out of two major influences. One was an economic potential. With enough payloads to be lofted regularly into space, it would eventually be cheaper to boost them into orbit by a system in which at least part of the boosting equipment returns to earth, like an airplane, to be used again and again. Furthermore, the shuttle is big and can take up a large number of payloads at the same time to share some common launch costs. The other influence was the growing pressure to reinject human beings into the space environment. The unsatisfied feeling remained with many that ultimately space must be added to the regions of the universe available to a human presence.

Thus, the shuttle project began. But the program planning and selling was dominated by a malady that has plagued many military programs; the idea was oversold and its costs were underestimated. The market, the need for frequent launchings of large weights, was pictured as larger and developing earlier than has actually happened. As has occurred so often with military programs, technological difficulties, time-to-completion and overall funding requirements were highly optimistically portrayed at the inception of the program. Schedules have slipped and more money has been required. The total payload to be carried has had to be revised downward, and the cost per pound to orbit payloads has had to be adjusted upward.

The lack of availability of the shuttle to boost military and commercial spacecraft into orbit on the required schedules has created problems and embarrassments. Some commercial American satellite projects that originally planned to use the shuttle have contracted instead to employ a new European-made, nonrecoverable booster that can meet the schedule and also offer a lower launching price.

The shuttle is still controversial. Some argue that if the market for payload orbiting had not been overestimated, the shuttle would have come along later. Proven nonrecoverable boosters then would have provided schedule reliability and least cost for required launchings in the interim. No American boosting business would have shifted to a foreign competitor. If the shuttle program had not been influenced by a premature interest in the highly speculative, supersized space structures of the distant future, the shuttle's physical design could have been simpler. The development might have required less time and money if the shuttle had been planned to be unmanned, and the cost to launch the majority of payloads for the next decade would be lower. That is what the critics say.

On the other hand, an American space program that assumes no role for human beings in space cannot be counted permanently as acceptable, since we cannot expect to anticipate all future needs. And there are always the Russians. They seem intent on maintaining a manned space station in orbit, and their program's progress will force continual comparisons with the U.S. space program. The USSR's specific goals in putting humans in space may be less significant than merely that such activities might be interpreted as indicating they are ahead. That possibility alone presses us to build a permanent U.S. space station.

THE PERMANENCE OF SPACE RESEARCH

Meanwhile, space research with unmanned spacecraft seems here to stay. We have, it is true, managed to learn a great deal about the universe from observations made on earth, a highly specialized, shielded, and insignificant little piece of it. However, by moving into outer space, we can get away from the handicaps of our atmosphere, magnetic field, emitted radiation, and gravitation, and expose some of nature's secrets totally alien to us in our isolated cubby-hole. But space is merely one frontier deserving investigation; in a well-ordered overall U.S. national research program, attention to it should compete with our efforts on numerous other knowledge boundaries. Still, we perform some research tasks ahead of others because for those chosen ones we happen to have the ready means, and the investigations can be done at reasonable cost. Other probings, just as ripe for important discoveries, may not be performable with the available tools. Proven spacecraft designs and deep-space communications networks are now extendable to perform most space research missions that deserve support.

Tracking the laws of the universe from regions of space never before available to scientists might well lead to discoveries we cannot possibly describe here because they are pure unknowns. Such possibilities might even have military applications. We cannot with great comfort allow some potential enemy to gain such new knowledge for its exclusive use. Space remains a new field in which to prospect; it is far too early to leave the finds to others by default.

The United States has exerted world leadership in exploring the solar system, and this has been one of our century's most inspiring scientific and engineering enterprises. If we do not go on exploiting the position America has attained, we will throw away a solid part of a major investment.

FREE ENTERPRISE IN SPACE

What are the potential benefits of business entrepreneurship in space, communications satellites in particular, since these come closest to being ready for delegation to the American private sector? An international telecommunications satellite agency, Intelsat, has been running quite smoothly for years and now has more than 100 nations as members. Intelsat provides more than 20,000 full telephone circuits, which represent about two-thirds of the world's total transoceanic telephone, telex, and data communications. The civilian market for communications satellites in the noncommunist world will require orbiting some 10 to 20 satellites per year during the 1980s. This activity, together with associated ground installations, adds up to several billion dollars annually for the R&D, satellite design, production of space hardware, and launch costs (as distinct from revenues from the services which those installations make possible and the investments in the associated ground networks). This is a large enough business that some half-dozen or more prime space contractors are likely to serve it—Japanese and West European as well as American. Some U.S. companies in the communications satellite field also can aspire to be suppliers of military communications satellites, which indirectly adds support to the commercial aspects. The worldwide civilian program alone, as it reaches the several-billion-dollar-a-year level, should support more than 100-million dollars of annual research and development expense. This should advance the art satisfactorily. A pattern of mutually reinforcing technology advance and revenue growth should develop and continue for the foreseeable future.

For some aspects of U.S. civilian commercial satellites, the government is now essentially out of the picture. Some argue that it could have been from the beginning. Even 20 years ago there were large communications companies in the United States (AT&T, IT&T, Western Union, RCA, and others) that could have justified adding the space dimension to their communications facilities. Even at that early time other electronic and computer companies should have suspected that communications satellites would directly affect their future growth plans. Because the government ICBM and Apollo programs had made available boosters, technology for design and production of the satellites themselves, and launch and tracking facilities, these companies needed to make only reasonable investments in the range of 100-to-200-million dollars each (with their net worths in the billions) to have begun developing satellite-based telephone and television broadcasting systems.

The government preempted this private approach, however, taking the initiative in establishing the Communications Satellite Corporation (Comsat), which since has become publicly owned; only a part of its Board of Directors is still appointed by the President. Left to themselves, the private communications corporations of that time would not have become involved with satellites until later. Had their leadership not lacked appreciation of the potential of space, imagination, and boldness, Comsat might never have been created. But because the industry possessed these shortcomings, the government saved time by dominating the beginnings of commercial communications satellite systems.

Although much of the initiative for civilian communications satellite projects in the United States now has passed into the hands of private corporations, the government still exerts a powerful influence. Every communications satellite system envisaged by the private sector must acquire operating frequencies and bandwidth in the limited radio spectrum, positional slots in synchronous orbit, and patterns of coverage of the earth below. All these are controlled by the U.S. government for American corporations. Moreover, what the U.S. government allows must be consistent with international agreements that divide up these privileges. Further, the satellites must be boosted into orbit, and the government still dominates the boosting of all spacecraft launched in the United States.

Even though launching technology relates to military boosters and such other associated classified apparatus as ICBMs, the U.S. government doubtless would permit independent, private booster activities. However, the art in boosters is constantly advancing as a result of the government's programs, and the resulting changes control what private groups will choose to do. The shuttle, for example, is altering the U.S. booster business greatly, and altering as well the potential for independent entrepreneurs to enter the field as private launch-vehicle providers. To make full use of the shuttle, the U.S. government will designate it to launch all government spacecraft and satellites in the future, military and nonmilitary. Had an American entrepreneur decided to develop a nonrecoverable booster, as the Europeans have done, other U.S. companies interested in orbiting commercial satellites clearly would have been prospective customers (since some such business has already gone to the European booster). However, that American booster company would have had difficulty getting any U.S. government business, even though its price to loft a satellite might be lower than the cost of using the shuttle for many of the government payloads now envisaged.

If in the future, in the commercial communications satellite field, booster equipment and the satellites that they put into orbit were totally available through free-enterprise activities, there would still be a need for launch and tracking facilities. The government owns the launch facilities in Florida and California. The tracking ranges involve a complex network of satellites, installations on U.S. soil, and installations in a number of other countries, arranged through intergovernmental negotiations. First created for the ICBM program and then extended for space programs, these installations were developed over a quarter of a century. To duplicate them would cost many billions of dollars, even if arrangements could be made again for suitable land in just the right places in the United States and around the world. A private group would find such a task prohibitively expensive. Moreover, the present facilities are adequate or readily extendable for handling all foreseeable traffic.

If launched from U.S. facilities, communications satellites, even those totally funded by the private sector for civilian applications, must fit their launch dates into a government program whose primary interest is military satellites and other government-funded spacecraft. Thus, even in the commercial satellite field, where the private sector is now closest to playing the lead, the U.S. government's involvement limits free enterprise.

Military satellites for command, control, communications, intelligence, and reconnaissance are not suitable for true free enterprise risk investment. The market has a single customer. Waiting for the government to decide which projects it wants is almost the whole of market research for would-be free-enterprise companies. The pattern is now well-established, and it is the same pattern found in all other classified military technology programs. Privately financed efforts, without government contracts, carried on to seek a head start in acquiring future military business in space, will probably be modest in scale.

The same applies to research space probes to Venus or Jupiter, to rendezvous with comets, or to provide platforms in space for sensing light, radio signals, or x-rays from distant parts of the universe. Such projects will remain under government sponsorship and will be totally financed by the government. Learning more about the laws of nature by observations in outer space is not an activity that an American corporation can justify funding with its own risk capital, even if extraordinarily long-range thinking characterizes its management.

As it establishes its priorities for pure research in space, as distinct from classified projects, the government should use the universities. Graduate studies leading to the higher technical degrees necessary for performing research and development are best undertaken in a university in which nature is being constantly explored. Space research should not be a difficult field for government-university relationships. Because the government is paying for space research, it must maintain an organization for making assignments and ensuring high-quality work. However, in recent years the government has made life difficult in research universities by "over-bureaucratizing" and over-administering. The attempt to ascertain that no research dollar is wasted can itself waste two dollars by over-documenting and setting excessive administrative rules.

THE GOVERNMENT/PRIVATE ROLES ASSIGNMENT

The roles assignment between the government and the private sector is much more complex and unsettled for certain other important applications of space technology than for communications satellites. One is airline navigation and traffic control through satellite systems. Here, no system to use space technology and aid the process of navigation and traffic control can exist without the participation—equivalent to a vote or a veto—of a very large number of semi-independent entities. These include the National Aeronautics and Space Administration (NASA), the airport operators, the airlines, the Federal Communications Commission (FCC), the Federal Aviation Administration (FAA), the Civil Aeronautics Board (CAB), the pilots' associations, the Army, Navy and Air Force, and the companies making the satellites, radars, airborne computer equipment, and ground-station apparatus. To make the system work, there must be rules requiring every airplane to carry appropriate equipment for cooperation with the system. Since foreign planes must operate in our environment, their governments must also be involved in establishing the specifications and in the installation of the system.

Clearly, there are numerous roles for units of the private sector with the required expertise. They should participate in studies, design proposals, research and development, and finally in the production of hardware and software to make the entire system operable. However, overall, this cannot be a free-enterprise project; the system responsibility is appropriately placed with the government. But there is no suitable government organization for space-based airline navigation and traffic control. No single entity has the responsibility to see this job through from beginning to end.

No law prevents the government from establishing an appropriate systems management unit, using private entities under contract to accomplish the necessary chores from systems engineering to hardware design. An integrated program was created for the huge ICBM system. But, lacking the organization to have the system designed and implemented, the United States has spent decades without yet having set up a satellite-based, airline navigation and traffic control system, even though it appears eminently sensible from the standpoint of economics, technology, traffic capacity, and safety. First, the government should recognize the organizational void and organize to fill it. Then the problems would be to enlist adequate support from the private sector and to do the job well. The latter steps, creating and implementing the system, will be difficult, but step number one has not even begun.

The government's role is better established in the weather field. Weather satellites have been operational for about 15 years and are now basic to weather forecasting, improving operating decisions in transportation, agriculture, fishing, and other fields. Here, the National Weather Service, with its ready-made infrastructure, was able to adjust for using and disseminating data from weather satellites. Amazingly, it was allowed to do so when the space dimension was opened. The World Meteorological Organization, in existence for many decades, was able to coordinate the interchange of data worldwide.

Landsat, however, a satellite to scan the earth's land resources, illustrates the puzzling policy questions to which applications of space technology can lead. Suppose a private American corporation were to orbit a Landsat as a free-enterprise project. This entrepreneuring company would process the data from probing the earth's surface and look for valuable information about mineral finds. By appropriate utilization of the information it would expect to reap financial rewards and realize a favorable return on its investment. The minimum investment required of such a postulated, risk-taking private corporation would be in the hundreds of millions of dollars. The company would need allocated frequency bands for communication, government approval of the orbital trajectories, and, for an appropriate fee, the use of government launch and tracking facilities.

These steps are easy to arrange. But other privileges the company would seek as it tries to exploit the acquired data would raise the question of whether it is proper to allow a private corporation to engage in Landsat activities. It would not be easy for the U.S. government to make this policy decision, but the problem would be compounded by the required international agreements in which the U.S. government would have to take the lead.

How would the risk-taker obtain revenues? If the data indicated the existence of a previously undiscovered valuable resource in a foreign country, should the company go to the government of that nation and offer to sell that information? How does the country thus propositioned even estimate the worth of the information unless it can analyze the data in detail and then perform on-site investigations of the scanned terrain? Perhaps the company simply should offer to sell either raw data or interpreted information to private groups or governments, who then would lease or buy the pertinent land or the rights to the resources on it.

But there is more. Observations made by a Landsat can disclose conditions of the earth's surface valuable for agricultural planning, controlling crop disease, finding and using water resources, dealing with mass pollution effects, anticipating and assessing the severity of flood conditions, and more. What should be done with these data, whose dissemination would be in the public interest and less likely to be adaptable to proprietary exploitation for profit? In fact, it is partly because of the potential gains for society that the U.S. government has already spent a billion dollars in the development of Landsat.

Its technology is not yet adequately developed and proven, and it is not clear that the system will uncover valuable finds. For a long time corporations dealing in petroleum and minerals may put their available financial resources behind more conventional discovery efforts. Only when (and if) the first important discoveries are made will it be the right time to consider how best to move the program along commercially. Meanwhile, the government will foot the bill for continued research and development, and this may produce valuable data both for public service (e.g., tracking crop disease) and resource discoveries (e.g., mineral ores).

Exactly how the United States should handle Landsat is unfinished business, although bills dealing with the problem recently have been introduced in Congress. The emphasis must be that this is an application of space technology with tremendous potential; certainly some of the potential is suitable for development by free enterprise. In America, how else can we really expect to find resources without involving free enterprise for stimulation, incentive, and risk-taking, with the anticipated appropriate rewards? Failure to use free enterprise in Landsat activities would be like expecting to discover oil only through government-sponsored searches. Yet the government must hold the chairmanship position, integrating the whole and parceling out appropriate pieces to the private sector.

Landsat is not the only space project that still needs assignment of appropriate government and free-enterprise roles. Ocean observation satellites, for example, display the conditions on and in the oceans, regions of the world largely beyond national jurisdiction. There is no existing institutional structure for operating an ocean satellite, yet the world might gain a great deal from a permanent system that acquires and distributes data about transport, fishing, and other ocean resources. Further, NASA is just beginning to grapple with commercializing the shuttle and otherwise converting it into an operational program once the R&D phase is completed. To what extent should free enterprise be involved in running the shuttle as a routine launcher of everyone's payloads for a fee?

It is perhaps understandable that the assigning of proper roles for the U.S. government and the American private sector in the commercial use of space is still incomplete. However, with space and its applications so clearly an area with powerful long-term economic and security interests for the nation, it is less forgivable that our approach to plans and policies about space has been on an intermittent, hop-and-jump, political, short-range basis for more than a decade. NASA does much hoping and planning, of course, but the United States does not have a plan for the next two decades—a real plan that lists goals and investment requirements, that has recognition and stature with the government's executive and legislative branches, that is totally clear to industry and the science and technology fraternity, and that has the long-term commitment of all these forces.

Comments

Aerospatiale
Société Nationale Industrielle
Paris, France

I would like to comment on five of the issues Simon Ramo has raised in his complete and excellent presentation: the military role of space; space and economy now and in the future; international competition and cooperation; space as a research frontier; and free enterprise and the government mission.

On the whole, I agree with Ramo on the military role of space. Beyond any doubt, if war should come, hostilities may spread to space. For that reason we must make every effort to reach international agreements so that the military use of space serves as a deterrent, preventing war. We should not try to forbid this kind of military activity in space; but we should improve the treaty signed in 1967, which forbade the launching of nuclear weapons into orbit.

On space and economy, Ramo quoted a value of a few dollars per hour per person as a kind of average for services involving a satellite. Such a value is perhaps low for long-distance telephone calls, but it appears very high when other activities are considered such as watching TV, airline navigation (per passenger), weather information, etc. On the other hand, I believe that many more than 100-million individuals will be effective users in ten years, so that, on a somewhat different basis, I can support figures on revenue and investment in the range of those given.

Another item is the limited number of available slots and radio frequency bands. In my opinion, Ramo is a little too pessimistic in this area. New frequency bands, such as the 20–30 gigahertz bands, will be available shortly, and it will be some time before there is a real shortage of either orbit slots or radio frequency bands. This is just one more reason to agree with Ramo when he indicates that negotiations between governments, or between governments and private corporations, constitute the initial bottleneck to progress in the commercial use of space. We, in Europe, have been involved in a number of such negotiations in the aerospace field, and we have found that such negotiations always create bottlenecks. But they have not stopped progress, and we can reasonably hope that they will not do so in the future.

The meaning of these reflections on economy is clear enough when they apply to such areas as telephony, television, airline navigation, meteorology, or data transmission. The horizon is a little bit hazier when we consider earth resources, and it is even more so for remote applications such as satellite-based solar power systems, manufacturing in space, or large space platforms. In all these areas, we must be cautious on feasibility and costs. At the recent 33rd Congress of the International Astronautical Federation (IAF), which took place in Paris in October 1982, there was a significant trend towards realism in these fields.

And applications of what Ramo calls "farther out" possibilities are still, and will for a long time remain, in the realm of research; economic considerations at this point are irrelevant. Nevertheless, we must work on them and, if I may reinterpret Ramo's analogy, as the automobile has not been invented we should strive to improve the breed of race horses.

Concerning the issue of international competition, I broadly agree with Ramo's analysis of the respective roles of man and machine in space. But I do not agree that the fact that some American customers have contracted to use a European launcher should be considered "sad events." I disagree not only as a European, but as a convinced supporter of international exchange and cooperation. It is quite true that America's enormous expenditures and pioneering efforts must receive a fair return. But how can anybody imagine that, in the long run, it will be better for the United States to have no one able to sell a launcher to them? That would mean not only the absence of motivating competition—Ramo took note of the role of competition in U.S. space progress—but also, in the long run, no solvent customers. In advanced technologies, as in many other fields, a two-way street is the best way for both partners to make progress.

Man-in-space is also one aspect of space as a research frontier. Biological studies and man's reactions in the space environment are bringing, and will continue to bring, a lot of knowledge to us. So is the case with solar system exploration. And I certainly agree with the idea that a lot of as yet unimaginable discoveries are possible.

Space is international by nature. This is well recognized when Ramo discusses free enterprise and the government mission. The communications satellite field is certainly the best example of free enterprise in space, but it needs world governmental intervention. Leaving aside purely domestic issues for the United States, I can only emphasize the role that governments will continue to play, even in this field. Nevertheless, free enterprise also has a significant role to play. We have established two companies, Arianespace and SPOT-IMAGE, in a way differing from Intelsat, but having the same philosophy. The purpose of these two companies is to operate in the free market, one for the launcher *Ariane* and the other for the data produced by SPOT, an earth-observation satellite. They were founded precisely because we thought it would not be right for the government agencies that developed these products to approach the marketplace. Their structure is not appropriate. A company that has to manage production and sale of such products must behave as a private corporation.

Another good example discussed by Ramo is airline navigation and traffic control through satellite systems. A number of private corporations have to participate in design, production, and operation of such systems. However, overall systems responsibility must be governmental. Ramo regrets that there is no appropriate systems management unit in the United States, although no law prevents it. Here I must add that willingness is necessary, too.

In the 1970s, an exceptional effort involving 20 countries, including the United States through its administration, resulted in a satellite-based navigation system proposal. But it encountered airline reluctance, and in 1977 it was finally killed by the U.S. Congress. This was a sad event for international cooperation.

So nothing is simple. Progress in the practical use of space requires at least four things: the willingness of partners, free enterprise activity, governmental support, and international cooperation. Planning this progress, as Ramo suggests, is a very difficult task. The area is so new that there will be, for a long time to come, much that is unexpected. Nevertheless, we may anticipate progress with confidence. Significant events at the last congress of the International Astronautical Federation encourage this hope. Its theme was "Space 2000," and many rather short-term, focused projects were discussed, for inevitably the year 2000 is entering the short-term picture. But this also reflected the trend towards realism, which is a symptom of good health. The proceedings of that congress include the following:

• analysis of possible systems for telecommunication in the band of 20/30 gigahertz;
• comparison of three earth resources satellite programs—Landsat of the United States, Earth Resources Satellite (ERS) of ESA, and SPOT of France—and studies on their developments;
• discussion of manned space stations, conducted in the presence of 18 astronauts from five countries—clear evidence of progress in international cooperation;
• sessions on cost reduction (a matter indicating realism, if anything); and, finally,
• a very good discussion of the necessity of developing international laws for space.

So, considering all that is happening, and even though we will have to endure many vicissitudes and adventures, I am sure that there will be rapid progress in the coming years, as there has been over the past 25 years.

Space is our future, the future for all of us here. We are part of the four billion astronauts, cosmonauts, or spationauts living on spacecraft Earth. Space is a common concern shared by all the world's people and so must provide opportunities for cooperation and peace.

Comments

EDWIN MANSFIELD
Department of Economics
University of Pennsylvania

I found Simon Ramo's paper very interesting and valuable. In my comments, I shall focus on the economic and commercial implications of the space program. In particular, I shall summarize very briefly and selectively the results of leading studies carried out to estimate the economic benefits from the space program.

First, I should describe the questions that these studies were designed to help answer. As Ramo pointed out in his paper, "when America launched its space program, it was an emotional reaction to the sputnik blitz," not the result of a carefully reasoned optimization of all our research and development opportunities. Nonetheless, it was not long before economists and policy makers were asking questions like these: How large are the effects of our investment in space R&D on the rate of productivity increase in the private sector of the economy? What is the social rate of return, or benefit-cost ratio, from this investment? How do the benefits from the space program compare with the opportunity costs?

None of these questions is easy to answer. To obtain information that might shed at least a limited amount of light on them, three types of studies have been conducted. First, there have been macroeconomic studies based on aggregate production functions. Perhaps the best-known of these studies was carried out by Chase Econometrics in 1975. Chase concluded that, over a decade, a one-dollar NASA expenditure would yield a cumulative return of 14 dollars in increased productive capability; this amounts to a 43 percent rate of return. Overall, Chase concluded that one-billion dollars sustained increase in NASA R&D per year would increase the gross national product by a cumulative 83-billion dollars by 1984.

Chase's study has been subjected to a variety of criticism. The General Accounting Office concluded that there was considerable instability in the equation used by Chase to explain the residual in the production function. The time period used in the study is short, relative to the long lags often involved in the introduction and diffusion of new technology. Moreover, an update of the Chase study, using data for 1956 to 1979, showed a statistically weak relationship between NASA R&D and changes in productivity. Despite these criticisms, however, macroeconomic investigations like the Chase study have been of considerable interest.

A second type of study has focused on NASA's records on programs designed to transfer technology or to encourage its transfer. For example, Henry Hertzfeld has presented data indicating that NASA granted 860 specific patent waivers between 1961 and 1975, and more than 20 percent of the relevant inventions have been commercialized. He also found that NASA has received about 3500 patents on both contractor-developed and employee-developed inventions, for which NASA has granted more than 500 licenses (on 200-plus patents) to both firms and individuals. However, only about 50 of these inventions were reported as commercialized by the beginning of 1979. In part, this may be due to NASA's basing its decision to patent on many considerations other than the invention's prospects for commercial success and to undocumented or unreported use of many of these inventions in the private sector.

A third type of study has focused on the economic benefits from individual innovations that stemmed from, or were hastened by, NASA R&D. Perhaps the best known study of this type was conducted by Mathematica in 1975. This study concluded that seven-billion dollars in benefits to the economy (by 1984) could be ascribed to NASA's involvement in stimulating the development of cryogenic insulation, gas turbine engines, integrated circuits, and NASTRAN (a computer program for analyzing structural properties of vehicles). This very significant stream of benefits stemming from only these four examples suggests that NASA's spending has had significant impacts on the private economy.

The case of cryogenic insulation illustrates the nature of the analysis. Mathematica estimated the increase in costs if the next best insulator, perlite, were used in place of cryogenic insulation in the transport of liquid hydrogen, liquid helium, and liquid nitrogen. To calculate the benefits due to NASA, Mathematica estimated that NASA had accelerated the benefits by a minimum of one year and probably by about five years. Based on the five-year figure and a 10 percent discount rate, the present value of the benefits from 1960 to 1983 is about one-billion dollars, which is hardly chicken feed.

In general, studies of the third type have received less criticism than the macroeconomic studies cited earlier. This is due partly to the fact that the microeconomic studies are based directly on data concerning the effects of specific innovations, not on the manipulation and partitioning of an aggregate residual, which is difficult to interpret. However, studies of the third type do have their own weaknesses. They too are frequently only blunt instruments, and in some important cases they cannot (or should not) be used at all.

Thus, many economic studies have focused almost exclusively on the short-term, often indirect, benefits of the space program. Economic forecasting being what it is, economists are probably wise to stay relatively close to the present. But it is, of course, quite possible that the great economic effects of the space program, some of which are discussed in Ramo's paper, may occur largely in the more distant future and may be quite different from the short-term effects. Many of the scenarios that have been suggested by technologists are striking indeed. If any of them comes to pass, the economic effects could dwarf those reported in the economic studies conducted to date.

Discussion

DICK PRESTON (Star Foundation)   An American, Edward Everett Hale, wrote a book on the brick moon—the first man, so far as I know, to suggest commercial use of a satellite for the purpose of navigation.

With Roger Chevalier now in charge of the International Astronautical Federation (IAF), what do you believe is the possibility for international cooperation on building an experimental space station for research and for international cooperation?

ROGER CHEVALIER   It is difficult for me to speak as president of the IAF on this subject. The purpose of the IAF is not to propose programs, but to assemble people to discuss what is being done in their countries and in their home organizations. Of course, I think that a joint program is certainly an interesting possibility. In the European Space Agency we have developed some programs among the European countries; the first one is a satellite launcher, *Ariane*. The second one is *Spacelab*, which is a part, in fact, of the shuttle program; so it is the beginning of cooperation between Europe and the United States. On the *Ariane* program, France is the leader; on *Spacelab*, Germany is the leader, and on the satellite called *MARECS* [Maritime European Communications Satellite], England is the leader. We prefer this solution, because it is more efficient to have one country or organization assume primary responsibility for a project.

SIMON RAMO   Americans should keep in mind that, in space and in technology and science in general, the combination of Europe and Japan includes more engineers and scientists now, and an equal ability to put resources behind each one, than does the United States.

79

Thus, in the future, we all will inevitably be sharing contributions to scientific development and utilization. As Roger Chevalier stated, competition has advantages. We have, in the past, had more extensive plans than we have today for cooperation in space. Unfortunately, the United States bowed out unceremoniously and suddenly because of budget reductions. Even if we assume the reductions were absolutely justified from the standpoint of the United States, they still left the Europeans holding the bag, after they made substantial investments in two expensive projects, especially as viewed by our European partners. One concerned Halley's comet, and another involved the examination of solar flares [the International Solar Polar Mission]. The United States portions were cancelled. I think it would be sensible, in the future, to conduct projects like space probes to Jupiter and Saturn cooperatively. These programs are big and expensive, and they will not bring immediate economic returns, but rather will further our understanding of the basic laws of nature. They are of value to the entire world. I think we will see and understand this, and necessity will force cooperation on us. We will have to improve our organization and, perhaps in the United States, learn how to make long-range commitments and stick to them, so that we will be regarded as a credible partner in this kind of research activity.

When I categorized American booster business going overseas as a sad event, I was referring to one very specific aspect of that situation—namely, that that business went by default. It would be one case if we looked at all that needs to be done in space, and the various nations and companies of the world looked at what they thought they could do best, sensed the markets, made their risk investments, tried to be best and first, and realized a return on their investments (with or without government involvement). It is another case when one country, the United States, sets out to do something and accidentally leaves a gap that could have been readily filled by the country's existing technology.

We are now entering an era in which, in view of the strength of western Europe and Japan in space, we should be competing in the classical sense. We should be picking out the projects that we individually think make more sense for us to do, with recognition of the fact that others are in the game as well. We will be right on some things; we will be wrong on others; and we may find ourselves sometimes duplicating each other's efforts. We may find we have picked out a nitch especially suited to someone else instead of ourselves and fail. And that will be true for others as well.

ROGER CHEVALIER   I do not consider the space shuttle and *Ariane* as competitors. They are, in fact, complementary; they do not have exactly the same goal. I am sure that, in the future, for reasons of cost, it will be necessary for certain small satellites to be launched by conventional launchers instead of the space shuttle.

JOHN GARDENEIR (U.S. Coast Guard)   Will anything be practical tomorrow that is not practical today? Our whole discussion on the practicality of space science focuses on the things that have been established for 10 years or so. There are other possibilities, such as asteroid mining, which certainly cannot be said definitely to be established as practical today. What are some of the new economic enterprises that might be developed in the next 10 to 15 years that are not practical today?

SIMON RAMO   There are, undoubtedly, applications in space that we do not even understand and have not even put our fingers on. But there are also some applications that have been specified and described. Most of these do not yet seem economically sensible.

On such things as manufacturing in space, for example, initial experiments have to come first. And these experiments are expensive. The means to carry them out will involve costly set-ups, often involving manned missions. Engaging in automatic manufacturing in space first requires developing a whole range of techniques that may not be of as much interest once we understand whether we can or cannot manufacture in space in a practical way.

We must also remember that at this time we still have a backlog of unfinished business. As I indicated in my paper, the airline navigation and traffic control system is an obvious case in which we have not made the necessary international arrangements to accomplish it. Here, I think, the American contribution has been especially inadequate. Maybe, as the rest of the noncommunist world becomes more active in space, that will push such things to completion. But to complete it, a great deal of work is yet to be done. In the case of earth resources satellites, the Europeans are coming into that field. The U.S. earth resources satellite project, Landsat, is still in the research stage. It has not yet shown that it will locate valuable resources, and Landsat is not yet ready for free enterprise. It is unlikely that an American entrepreneur, a private corporation, no matter how large, is going to want to take a billion-dollar risk to see if it can further develop earth resources satellites. When we note that Exxon, a corporation in the 100-billion-dollar category, has decided not to take the risk of developing the capacity to obtain liquid fuel from shale, we can see that such developments will still depend on government budgets. In this particular period, that kind of government investment is at a low ebb.

We have a lot of unfinished business, promising projects that need to be investigated. We have only proceeded 10 or 20 percent along these lines of investigation, so I do not feel compelled to lengthen the list, particularly when the additional items look even more expensive and even more speculative than those already in process. In a few years, however, the whole situation may change.

SECTION 3

Science and Space

Introduction

GERALD HOLTON
Department of Physics
Harvard University

What lies behind our desire to consider the quarter of a century since the launch of mankind's first probe into orbit? It is, perhaps, the feeling that the time is ripe for an informed look in each direction—some speculation about the future and a first assessment of the recent past. Some may feel, as I do, that this leap into space witnessed in our time will remain part of the permanent memory of mankind, alongside the historic memory of the great journeys of adventure and discovery that formerly found expression in epic form. People in the distant future may, in their own way and using their own media, be describing our accomplishment of transcending our physical dependence on the earth somewhat as we still are singing Homer's song to relive the voyage of Odysseus beyond the boundaries of the classical world.

I cannot help wondering what history will say about the topic of this section, "Science and Space." I venture two brief speculations. The first is that, in retrospect, the intentional exploration of the solar system, by means of those remarkable earth-launched physical instruments, was prepared for by, and dependent on, a series of equally daring, mental launchings into space. Science and space have in fact been Siamese twins from the start: Space has been the foremost laboratory of the scientific imagination—from the pre-Socratics who toyed with the question of the limits of space, to Aristotle and his followers for whom the cosmos was not only finite but relatively small, to Kepler who could envisage something like the law of conservation of momentum by thinking about mutually attracting and colliding bodies in space, to Galileo for whom space was not yet Euclidean but warped, and on to Newton and the modern period. In a sense, which will become more obvious as time passes, the space age really started not with *Sputnik I* but with those early explorers of the mind's own spaces, whose conceptions were the necessary preparation for the launching of the hardware.

There are several candidates for a designation of the father of the space age. My own preference is a philosopher, mathematician, cosmologist, and cardinal of the church, Nicholas of Cusa. Appropriately, a crater on the moon has been named after him. A good description of his work is in Alexander Koyré's great book, *From the Closed World to the Infinite Universe*. Nicholas of Cusa, who lived from 1401 to 1464, was one of those who tried to break out of the geocentric, anthropocentric, finite, and hierarchically sequenced world of antiquity, a world bounded by the walls of the heavenly spheres. He glimpsed the dizzying potential of space and entertained a very different universe: open, unbounded, without natural subordination of any one part to any other, filled with identical laws and with essentially interchangeable components. Technically, his step is called the "infinitization of the cosmos," an idea so new then that it was ignored by Nicholas of Cusa's contemporary, Copernicus, who thought the world was contained within a sphere of about 20,000 earth radii.

But Nicholas of Cusa saw the consequences of his vision: In an immeasurable universe, where there is no limiting point or center, all motion is relative, and the earth and all other bodies may be considered in motion. The earth then joins the ranks of the noble stars, and he even imagined that the stars may also be endowed with life forms. Most of his readers recoiled in horror and vertigo, except Giordano Bruno, who embraced these ideas, and who, by being burned at the stake in 1600 for such heresies, became (so to speak) the first space casualty. Thereafter, however, Nicholas' ideas became more and more influential.

Nicholas of Cusa was a prominent person, but we know all too little about him. Though we happily have his book with the modest title *On Learned Ignorance*, which I like to think started the space age

542 years ago, some of the most adventurous of his scientific-philosophical writings have been lost to history.

My first speculation, then, is my thought of the real father of the space age, which brings me to the second point: Who, in the long run, will tell *our* story? Will the future students of our accomplishments have reliable information, more reliable than we have about our predecessors? Who is now concerned with preparing accounts that can withstand the scrutiny of the ages to come? Who is saving the database, the less obvious documentation of successes and failures? Who is conducting the oral history of the pioneers? Are they able to handle the science, the technology, and the industrial and administrative components of modern space achievements?

There are a few who can. They are the historians of science and technology. On the members of that young profession we shall have to rely for the preservation of the record and for the assessment and the authentication of what really has been happening during the recent heroic period. We are lucky that a few such people in the United States dedicate their lives to such scholarship. There are also such scholars outside the USA, particularly in the USSR, which includes the historians of science and technology in its Academy of Sciences and which has a well-financed Academy Institute in this field. In France there is also a vigorously growing presence in the history of science and technology. But altogether, the numbers of professionals are few, and in the United States their support and the infrastructure of their professional societies are now under severe and increasing constraints. So I ask myself: Who will be the future Homers to sing of our time, and where will they get their information?

This symposium is, of course, a good step in the right direction, the more so if we can bring out novel or unusual points of view for fruitful debate. Our contributors are well selected for that purpose. Freeman Dyson, born in 1923 in Crowthorne, England, earned his Bachelor of Science degree from Cambridge, where he later became a Fellow of Trinity College. He first came to this country as a Commonwealth Fellow at Cornell and Princeton universities. In 1951 he emigrated to the United States permanently, becoming a professor at Cornell that year and then moving to the Institute for Advanced

Study in Princeton in 1953. Dyson became a naturalized U.S. citizen in 1957, and in 1962 and 1963 he was chairman of the Federation of American Scientists. He is a Fellow of the Royal Society, a Fellow of the American Physical Society, and a member of the National Academy of Sciences. Among his publications is, of course, that delightful book with the unsettling title, *Disturbing the Universe* (1979).

Freeman Dyson has been a consultant for NASA, and he is a member of the Space Science Board of the National Academy of Sciences. Among his countless honors he has received the Heineman Prize of the American Institute of Physics (1965), the Lorentz Medal from the Netherlands Academy (1966), the Hughes Medal from the Royal Society (1968), the Max Planck Medal from the German Physical Society (1969), the J. Robert Oppenheimer Memorial Prize from the Center for Theoretical Studies (1970), the Harvey Prize from the Israel Institute of Technology (1977), and the 1981 Wolf Prize. When you ask him what he does, he will tell you he is a mathematical physicist who is interested in astronomy. But we know better. He is interested in any question that may have a scientific solution.

The first commentator on Dyson's paper is Hendrik van de Hulst. Born eight days after the end of World War I in Utrecht, the Netherlands, van de Hulst felt war's impact in his student days, when his teacher, M. Minnaert, was dragged off to a World War II hostage camp. Left behind, much like Kepler in similar distress centuries earlier, the student turned his thoughts to astronomy. In 1944, he completed his seminal paper predicting the 21-centimeter line for radio astronomy. While he was wondering whether a transition arising from the hyperfine splitting of the lines in the hydrogen spectrum would be easily detected, his calculations revealed, to his astonishment, that this transition, which occurs in the most abundant species of matter in the universe, manifests itself in radio waves of an almost ideal frequency. "Isn't now the whole sky gleaming at this wavelength?" is how he expressed the implications of his work—a vision worthy of Ezekiel.

Van de Hulst received his Ph.D. from the University of Utrecht in 1946, and six years later he became a professor at Leyden. He has been a visiting professor several times, including 1951 at Harvard. There he was in time to see Edward Purcell and his students sticking a radio horn out of a window in Lyman Laboratory to test—and prove—the existence of the 21-centimeter line. In 1958 van de Hulst was elected the first chairman of the International Committee on Space Research (COSPAR), and he remained on that committee until 1963. During the same period he also undertook the long-term responsibility of chairing the Netherlands Committee for Geophysics and Space Research. Among his awards are the Eddington Medal of the Royal Astronomical Society, the Draper Medal of the U.S. National Academy of Sciences, and the Rumford Medal of the Royal Society of London. He has written several books and currently lists among his research interests interstellar matter, the solar corona, and the scattering of light in planetary atmospheres.

The final commentator is Gerald J. Wasserburg. Born in 1927 in New Brunswick, New Jersey, he served in the U.S. Army during World War II. He obtained his Bachelor of Science degree in Physics from the University of Chicago and remained there to receive his Ph.D. in Geology in 1954. Since 1962 he has been at the California Institute of Technology, where he is John D. MacArthur Professor of Geology and Geophysics.

Among Wasserburg's many awards are two Distinguished Public Service Medals from the National Aeronautics and Space Administration. He has received the Arthur L. Day Medal of the Geological Society of America and recently accepted the Arthur L. Day prize and lectureship from the National Academy of Sciences and a Smithsonian Institution Regents Fellowship. He is a member of the National Academy, a Fellow of the American Geophysical Union, the American Academy of Arts and Sciences, and the Geological Society of America. Wasserburg's research involves the application of the methods of chemical physics to geological problems, the measurement of absolute geological times, solar-system and planetary time scales, and the time scales associated with nucleosynthesis as inferred from isotopic studies. His professional interests include a passionate concern for the future of scientific excellence in this country.

Science and Space

FREEMAN J. DYSON
Institute for Advanced Study
Princeton, New Jersey

TWO POINTS OF VIEW

Science has never been the main driving force of the space program, and the space program has never been the main driving force of science. That is as it should be. Science and space each has its own objectives and grand designs, independent of the other. Science is at its most creative when it can see a world in a grain of sand and a heaven in a wildflower. Heavy hardware and big machines are also a part of science, but not the most important part. Conversely, the space program is at its most creative when it is a human adventure—brave men daring to ride a moon-buggy over the foothills of the lunar Apennines to the brink of the Hadley Rille. Precise observations and dating of moon rocks are also a part of space exploration, but not the most important part.

The main driving forces of the space program have been political, military, and commercial rather than scientific. If we measure the size of programs by total effort and budgetary outlay, then roughly 10 percent of the space program is science and roughly 10 percent of the science program is space. Nevertheless, the 10-percent area of overlap between science and space is of vital importance to both parties, and since I am a scientist I shall concentrate on this area of overlap in my remarks. From a survey of the highlights of space science during the last 25 years I shall try to derive some useful lessons for the future. My conclusions are necessarily personal, based

90

on limited experience and partial knowledge. Fortunately, there is space in this volume for other opinions.

There are two ways to look at space science: from the space side or from the science side. If one comes from the space side, it is natural to adopt a mission-oriented approach, measuring success and failure by missions done and not done. An examination of the last 25 years from such a point of view reveals a large number of splendid successes, a smaller number of sadly missed opportunities, and a very few outright failures. This is the point of view of the space professionals, and it is also the point of view of the general public insofar as the general public is interested in space science at all. I do not claim that the mission-oriented approach is wrong. But it is not the whole story.

Since I come from the science side, I look at space science with a science-oriented approach, which measures the success and failure of missions by looking at the quality of their scientific output. This approach sees space science as embedded in a wider context of ground-based science, and asks of each mission not merely the easy question—Did it work?—but also the more difficult questions: So what? What did we really learn? Was that the right thing to observe? Was that the quickest, or the cheapest, or the most effective way to make the observation? The science-oriented approach does not believe in pass-fail grading. Scientifically speaking, there can be total failure of a mission, but never total success. A successful mission will raise new questions as often as it answers old ones.

I like to deal in particular instances rather than in generalities, so I will begin my discussion with a concrete example—a successful space mission that I know something about because it was conceived and operated in Princeton. The orbiting ultraviolet telescope called *Copernicus* was launched in 1972, just in time to celebrate Copernicus' 500th birthday. From a mission-oriented point of view, *Copernicus* was a brilliant success. It did exactly what it was designed to do, taking high-resolution ultraviolet spectra of hot stars and measuring absorption lines produced by atoms and ions in interstellar gas. It was supposed to last for one year and actually lasted eight, producing data, year after year, until it finally died of old age. People at Princeton are still studying and publishing papers based upon the data.

The Princeton astronomers are justly proud of their *Copernicus*; they invented it, designed it, fought for it, used it, and nursed it through its declining years—a remarkable achievement for a small university department and for the NASA Office of Space Science and Applications.

But from the science-oriented point of view, the picture is more complicated. The original idea of *Copernicus* arose in the 1950s in the mind of Lyman Spitzer, who was, and still is, a pioneer in exploring the nature and distribution of interstellar gas in our galaxy. At that time the main evidence for the chemistry of interstellar gas came from narrow absorption lines of sodium and calcium seen in the optical spectra of certain stars. Sodium and calcium are the only elements that have absorption lines in the part of the optical spectrum seen with ground-based telescopes. But they are minor constituents of the gas, so the sodium and calcium lines do not give good information about the behavior of the gas in general. The majority of atoms in the gas belong to the common elements hydrogen, carbon, nitrogen, and oxygen, which have absorption lines only in the far ultraviolet. So Spitzer decided it would be a good idea to put into orbit a far-ultraviolet telescope able to record and measure accurately the absorption lines of the abundant elements in the interstellar gas. NASA agreed, and Spitzer's telescope was approved in 1960 as the third in the series of Orbiting Astronomical Observatories. The contract for its construction was signed in 1962, with launch originally scheduled for 1965.

For various reasons, partly technical and partly political, the launch of *Copernicus* was delayed seven years, so that a telescope designed to answer the scientific questions of the 1950s was not launched until the 1970s. But between the design and the launch of *Copernicus* a revolution occurred in radio astronomy. Radio astronomers observing from the ground learned how to see the interstellar gas with millimeter waves, which answered the main questions about the chemical composition of the gas. Millimeter-wave telescopes on the ground did a large part of the job *Copernicus* was designed to do, more quickly, more cheaply, more comprehensively. This does not mean that *Copernicus* was scientifically useless. Its observations complemented the millimeter-wave observations nicely, giving information especially about the dilute high-temperature component of interstellar gas, which is invisible to radio-telescopes.

Thus, *Copernicus* is far from being a scientific failure. But it was not the instrument astronomers would have chosen to answer the exciting scientific questions of the 1970s. By then we needed an ultraviolet telescope with which we could study the newly discovered x-ray sources, quasars, and other mysterious objects in which violent dynamic processes are occurring. All the newly discovered objects are faint, and *Copernicus*—because we had irrevocably chosen to sacrifice light-gathering power for the sake of high spectral resolution— could not see faint objects. The only time *Copernicus* had a chance to look at an exciting new object was when Nova Cygni flashed in the northern sky for a few nights in August and September of 1975.

Of course *Copernicus* was only one of many successful missions in the scientific part of the space program. I mention it in detail here because I believe it illustrates a general problem that recurs again and again in the history of space science—that is, the mismatch in time-scale between science and space missions. The cutting edge of science moves rapidly. New discoveries and new ideas often turn whole fields of science upside-down within a few years. The discovery of pulsars in 1967 burst on the astronomical scene as suddenly and unexpectedly as did Nova Cygni, and transformed, within a year, the way we thought about the late phases of stellar evolution.

The effect of such discoveries is to change the priorities of science, to change the questions we want to answer. Every young scientist's dream is to be able to say what the 19-year-old mathematical genius Evariste Galois said in 1830, "I have carried out researches which will halt many savants in theirs." Science must always be ready to halt and switch its objectives at short notice. Therefore, the tools of science should be versatile and flexible. But flexibility and versatility are hard to achieve in space missions. In the space program, plans for missions and designs of instruments tend to be frozen many years in advance. *Copernicus*, with a 12–year interval between design and launch, was perhaps an extreme case. But intervals of 8 and 10 years are not uncommon. In most major space missions, the instruments were designed to answer the questions that seemed important to scientists a decade earlier. The bigger and more ambitious the mission, the more difficult it is to reconcile its time-scale with that of science. Space science begins to look more and more like a two-horse shay with a cart-horse and a race-horse harnessed together.

In science, therefore, quality is more important than quantity. Bigger budgets and grander missions do not necessarily lead to better science. If we want to conduct good science in space, we must preserve our flexibility and have available a wide variety of missions and instruments, so that we can move quickly to take advantage of unexpected opportunities. The most important discoveries are those that could not have been planned in advance.

HISTORY

What should the space program do to recapture its lost youth? Before answering, I will look at the history of the first 25 years of the space program from a science-oriented point of view. Looking at the past provides an opportunity to learn from mistakes how to do better in the future, and to learn from successes how not to do worse in the future.

The 25 years since *Sputnik I* divide themselves conveniently into two periods: Apollo and post-Apollo. The Apollo period ends with the departure of Harrison Schmitt and Eugene Cernan from the Moon in December 1972. It is a particularly instructive period, because we can see clearly, with the benefit of hindsight, which parts of the space enterprise in those years were scientifically the most productive. At that time there was, in fact, a very strong negative correlation between budgetary input and scientific output, which was neither planned nor expected. It just happened because science is unpredictable. The most expensive missions produced the least significant science, and the cheapest missions produced the most exciting science.

The space program of the Apollo period included three main types of exploratory mission: manned missions culminating with Apollo, unmanned planetary missions culminating with the Mars and Mercury *Mariners*, and a series of x-ray sounding-rocket missions culminating with the launch of the first x-ray satellite, *Uhuru*. The costs of the Apollo missions, the *Mariner* missions, and the x-ray missions were very roughly in the ratio of 100:10:1. We cannot attach numerical values to the scientific results of the various missions; to some extent these are a matter of personal taste. Nevertheless, I am prepared to say unequivocally that the beginning of x-ray astronomy, opening a new window on the universe and revealing the existence of several new classes of astronomical object, was the most important single scientific fruit of the whole space program. The newly discovered x-ray sources gave an entirely fresh picture of the universe, dominated by violent events, explosions, shocks, and rapidly varying dynamic processes. X-ray observations finally demolished the ancient Aristotelian view of the celestial universe as a serene region, populated by perfect objects moving in eternal peace and quietness. The old quiescent universe of Aristotle, which had survived the intellectual revolutions associated with the names of Copernicus, Newton, and Einstein essentially intact, disappeared forever as soon as the x-ray telescopes went to work. And the new universe of collapsed objects and cataclysmic violence originated in the cheap little sounding-rockets of the sixties, popping up out of the earth's atmosphere and observing the x-ray sky for only a few minutes before they fell back down. The most brilliant achievement of the sounding-rocket era was Herbert Friedman's 1964 measurement of the angular size of the x-ray source in the Crab Nebula using the moon as an occulting disc. Yet the cost of x-ray astronomy in the Apollo period was much less than one percent of the total budget for space.

The manned missions, which absorbed the bulk of the space budget in those days, yielded a harvest of solid scientific information about the moon. Samples of various types of moon rock were brought home, analyzed, and dated. The stratigraphy of the moon was clarified and its early history elucidated; its seismic and magnetic characteristics were measured. All this was good science, but it was not great science. For science to be great it must involve surprises; it must bring discoveries of things nobody had expected or imagined. There were no surprises on the moon comparable with the x-ray burst sources or with the x-ray binary sources, which gave us the first evidence of the actual existence of black holes in our galaxy.

Everything discovered on the moon could be explained in terms of conventional physics and chemistry.

The unmanned planetary missions of the Apollo period were intermediate, both in cost and in scientific importance, between the manned missions and the sounding-rockets. They were less costly than Apollo and less exciting scientifically than x-ray astronomy. Perhaps the most exciting aspect of the planetary missions was their technical brilliance. The triple encounter of *Mariner 10* with the planet Mercury, a game of celestial billiards invented by Guiseppe Colombo, demonstrated the spectacular skill and daring of the engineers at the Jet Propulsion Laboratory. The *Mariner* missions gave us, in addition, some beautiful scientific surprises—high temperatures and pressures and the absence of water in the atmosphere of Venus, giant volcanoes and canyons, ancient craters, and the absence of canals on Mars— but the surprises were not of such a magnitude as to cause a scientific revolution. The newly discovered features of Mars and Venus were mysterious but not wholly unintelligible. The *Mariner* observations were a big step forward in the understanding of the planets; they were not the birth of a new science.

So much for the Apollo period. If space exploration had stopped at the end of 1972, we might have deduced from that history that a simple mathematical law governs the scientific effectiveness of missions in space, namely, that the scientific output varies inversely with the financial input. If this law held universally, the administration of space science programs would be a simple matter: Just cut the budget and watch the science improve. But this simple managerial method does not always work as it should. The history of space science in the post-Apollo period shows a more complicated pattern.

In the 1970s we again had three programs continuing the work begun in the 1960s: the *Skylab* and shuttle missions taking over from Apollo, the *Viking* and *Voyager* missions taking over from the *Mariners*, and the *Einstein* x-ray observatory taking over from *Uhuru*. Once again, the x-ray observations were first in scientific importance. During its short operational life, *Einstein* poured out a steady stream of revolutionary discoveries, including the discovery of x-ray variability of quasars on a time-scale of hours. The rapid variation of quasars implies that some of these objects have a switch, which can turn the energy output of 10-billion suns on and off within 100 minutes. The x-ray telescope allowed us for the first time to penetrate close to the central core of the mysterious engines that drive these most violent objects in the universe. Thus, in the 1970s the x-ray discoveries were again of greater fundamental importance than the planetary discoveries, even though the *Viking* and *Voyager* missions yielded a wealth of new scientific information as well as pictures of incomparable beauty. And again, the planetary missions outstripped *Skylab* in scientific value. However, in the 1970s, unlike the '60s, there was no longer a factor-of-ten difference in costs among the three types of mission; they all had become comparably expensive. *Einstein* was a little cheaper, not enormously cheaper, than *Skylab*. By the end of the 1970s, we could no longer be as confident that the smallest and cheapest parts of the space program were scientifically the best. All parts of the program, irrespective of their scientific merit, had come to be dominated by large and expensive missions, and the program thereby lost the flexibility that the small missions of the Apollo period had kept alive.

I am not saying that big expensive missions are unnecessary or undesirable. On the contrary, the *Voyager* and *Einstein* missions were great scientific achievements, and all scientists should be grateful to the United States taxpayers for the generous funding that made them possible. Big expensive missions have an essential role to play in space exploration, but a space science program needs also to put a substantial fraction of its effort into small missions if it is to keep pace with the shifting needs of science. A program dominated by large missions is in danger of losing its scientific vigor, even if it maintains its political support.

Large missions have two outstanding defects, which are apt to lead to scientific trouble. The first I already mentioned: the long lead-times, which make large missions inflexible and unable to respond to new ideas. The second is the tendency of big missions to become one-of-a-kind. This defect is related to the political climate within which large missions must be presented to the government and to the public. In order to secure funding for a large scientific mission, the proponents must talk about the important scientific problems that that mission, by itself, will solve. Then, to stay honest, they must conform the design of the mission to their promises. The mission then becomes a one-shot affair, designed and announced to the public as the final answer to some big scientific question. The most "unhappy" example of this one-shot syndrome is the *Viking* mission, which was forced by the political circumstances of its origin to accept the impossible scientific task of deciding, all by itself, whether or not life exists on Mars. If one looks in detail at the *Viking* experiments, it is difficult to imagine any combination of results that would definitely have proved or disproved the existence of life on Mars, unless we had been lucky enough to find a cactus bush or an armadillo sitting immediately in front of its television cameras.

One-shot missions are simply not a good way to do science. If we want to investigate seriously the question of life on Mars, the best way would be to plan a regular series of Mars missions, each one far less ambitious and elaborate than *Viking*, so that we could learn from the results of one mission the right questions to ask with the next one. We could also learn how to avoid the mistakes of one mission on the next. In almost any field of space science, whether it be exploring planets or galaxies or earth, a series of modest missions is more likely to produce important discoveries than a single big spectacular.

The baleful effects of the one-shot syndrome are clear not only in the *Viking* mission but also in x-ray and optical astronomy. The *Einstein* x-ray observatory was magnificently productive while it lasted, but it was a one-shot performance with no follow-on. We must now wait many years before another mission can be launched to answer the new questions that *Einstein* raised. We would have been far better off scientifically with two or three small *Einsteins* in sequence, instead of one big one. I must also confess that I am uneasy about the scientific justification of *Space Telescope*. It is of course the grossest heresy or treason for a scientist to express any misgivings about *Space Telescope*. *Space Telescope* is the grand centerpiece of the whole space science program for the 1980s. It will undoubtedly be a splendid instrument and will extend massively the boundaries of optical astronomy. But I am uneasy when I hear its proponents speak of it as *the* telescope, one destined to dominate optical astronomy for the remainder of this century. We scientists must be grateful for anything we can get, and I am indeed grateful to NASA for providing us with a large telescope in space. But, in all honesty, I believe that *Space Telescope* is a basket with too many eggs riding in it. It would have been much better for astronomy if we had had several one-meter space telescopes to test the instrumentation and see how the sky looks at one-tenth-of-a-second-of-arc resolution, instead of being stuck with a single one-shot 2.4-meter telescope for the rest of the century. Perhaps I am being unduly pessimistic. *Space Telescope* can hardly fail to make big discoveries when it comes into service in the 1980s. But I have a sneaking fear that it may end up in the 1990s, as *Copernicus* did in the 1970s, a glorious technical success but scientifically 20 years behind the times.

PLANS FOR THE 1980s AND 1990s

Space Telescope is only one item in the plans for space science in the coming decade. The plans are subject to great political uncertainties. A large number of ambitious missions have been proposed and recommended by various committees of distinguished scientists, but few have been officially approved and funded.

Three missions, in particular, have been approved and scheduled to fly in the 1980s: *Space Telescope*, *Galileo*, and *Hipparcos*. *Space Telescope* and *Galileo* are shuttle missions, both stretching the limits of budgetary and political feasibility. Both have been subject to long delays and technical uncertainties associated with difficulties in the development of the shuttle. *Space Telescope* is a general-purpose optical instrument designed to give images about twenty times sharper than the best images obtainable from ground-based telescopes. It will explore the fine details of selected objects, mostly very dim and distant objects that cannot be effectively observed from the ground.

Galileo is a planetary mission, which will do for Jupiter what *Viking* did for Mars, sending probes deep into Jupiter's atmosphere and providing fairly complete photographic coverage of Jupiter and its satellites. Like *Space Telescope*, *Galileo* is a one-of-a-kind mission. No further large missions to Jupiter are planned before the end of the century. If, as is likely, *Galileo* raises important new questions, we will have to wait a long time for a chance to find the answers.

Hipparcos is a bird of an entirely different color. In the first place, it is not a NASA project. It belongs to the European Space Agency, having been invented and originally proposed by Professor Lacroute of Dijon, France. In the second place, it is independent of the shuttle, being small enough to be placed into a geostationary orbit by the French *Ariane 1* launch system. In the third place, it is cheap enough to be the first of a series. If the first *Hipparcos* mission works well, it will be easy to launch follow-on missions to give us higher precision or more extensive coverage. If the first one fails, it will not be a major disaster.

Hipparcos is an astrometric satellite, designed to do nothing else but measure very accurately the angular positions of stars in the sky. It will give positions about 10 times more accurate than those measured by ground-based telescopes. This sounds like a modest and unrevolutionary objective, but, in fact, the improvement of positional accuracy is of central importance to astronomy. If we can improve the accuracy of angular position measurement by a factor of 10, we increase by a factor of 10 the distance out to which we can determine the distances of stars by the method of parallaxes. Thus, we increase by a factor of 1000 the sample of stars whose distances we can reliably measure. In other words, the improved positional accuracy will give us a stereoscopic, three-dimensional view of several hundred-thousand stars, instead of the few hundred that lie close enough to have parallaxes measurable accurately from the ground. When a star's distance is known its absolute brightness is also known, and the absolute brightness is the most important quantity determining the structure and life-history of a star.

The tenfold extension of our range of stereoscopic vision will have qualitative as well as quantitative importance. The few hundred stars whose distances we can now measure accurately are a random sample of those that happen to lie close to the earth; they are almost all dwarf main-sequence stars of the commonest types, giving little information about interesting phases of stellar evolution. When we have a sample of several hundred-thousand stars of known distance and known absolute brightness, it will include many rare types, such as variable stars of different kinds, which we are observing at crucial transient phases of their lives. In this and other ways, the data from *Hipparcos* will give us a wealth of new information about the constitution and evolution of stars and about the dynamical behavior of our galaxy. The *Hipparcos* mission also includes a completely automated data-processing system on the ground, so that star positions will not be measured laboriously one at a time but will be computed wholesale in batches of 100,000. The data-processing system will be a more revolutionary improvement of the state of the art of astrometry than the satellite itself.

I hope I am not exaggerating the virtues of *Hipparcos*. I do not wish to embarrass my European friends by giving their brainchild more praise than it deserves. But two facts about *Hipparcos* seem to me to be of fundamental importance. First, it is the first time since *Sputnik I* in 1957 that a major new development in space has come from outside the United States. Second, it is the first time since the days of *Uhuru* that a major new development has come from a small and relatively cheap mission, one that can be repeated and further developed without putting excessive strain on launch facilities and budgets. These two facts are a good augury for the future. I believe that space science will flourish only if we can move away from grand one-shot missions like *Space Telescope* and *Galileo* toward smaller and more flexible missions in the style of *Hipparcos*. And *Hipparcos* is probably only the first of many good ideas that will come from the space science programs of Europe and Japan, giving us the competitive stimulus that the Soviet space science program once promised but has dismally failed to maintain.

INTO THE WILD BLUE YONDER

I have been speaking very critically of the United States space science program as it now exists. The program is not only in political trouble but also in scientific trouble, and I am deliberately emphasizing the trouble we are in. It would be a waste of time for me to tell you about all the things that we have been doing right; the way to a better future is not to deny mistakes but to learn from them. We shall not succeed in overcoming our political difficulties unless we also admit and deal honestly with our scientific shortcomings.

On a more positive note, I would now like to describe some of the great opportunities that still lie open for doing first-rate science in space—some possible directions in which we may look for a renaissance, a new flowering of space science. Some of these new directions could be started today; others require technology that may take 10 or 20 years to develop. I will try not to be too imaginative; that is, I will talk only about possibilities that have some chance of being realized before the next anniversary symposium in the year 2007.

Four projects could be undertaken in the fairly near future. We would be starting the first one right now if we had our scientific priorities straight, namely, the orbiting VLBI observatory designed by Bernie Burke at the Massachusetts Institute of Technology (MIT) and by Bob Preston and his colleagues at the Jet Propulsion Laboratory. VLBI means very-long-baseline interferometry. During the last 20 years the techniques of VLBI have been developed and refined by radio astronomers working with ground-based telescopes. VLBI is one of the most spectacular successes in the history of astronomy; it already allows us to observe distant radio sources with angular resolution 100 times better than the best *Space Telescope* will be able to do. The performance and versatility of ground-based VLBI systems are still improving rapidly and leaving the optical astronomers further and further behind.

The secret of success in science is to put your money quickly on the winning horse. The orbiting VLBI observatory of Burke and Preston would be a single radio antenna of modest size, orbiting the earth in an elongated elliptical orbit and adding its signals to the existing network of VLBI telescopes on the ground. A similar orbiting VLBI antenna has been proposed and designed by Nikolai Kardashev in the Soviet Union. The addition of a single space antenna would improve the capabilities of the ground-based system by a factor of about 10, and that would not be the end of the story. Like *Hipparcos*, the orbiting VLBI observatory is a comparatively small, cheap mission—small enough to ride piggyback into orbit with some higher priority spacecraft and cheap enough to be repeated if it works well.

Within 10 years we could have a network of orbiting VLBI observatories in a variety of orbits, pushing the angular resolution of radio-astronomical observation toward the ultimate limits set by the lumpiness of the interstellar plasma. And all on a learn-as-you-go basis, free from the rigidities of a one-shot mission.

The second item on my list of good bets for the future is astrometric spacecraft, using and extending the technology of *Hipparcos*. There is no reason why we should leave this area of science entirely to the Europeans. It would be extremely rewarding to extend the range of stereoscopic vision of optical astronomy still further, either by improving the precision of the *Hipparcos* optical system or by launching *Hipparcos* spacecraft away from the earth as far as Saturn to obtain parallaxes on a 10-times longer baseline.

The third project is optical interferometry, which should be as spectacularly successful as radio interferometry has been on the ground. Optical interferometry does not require large telescopes or large rigid structures. Early missions could be quite modest, with baselines of a few tens of meters and telescope apertures of a few inches. This would enable us to map the optical structure of bright objects with angular resolution 10 times better than *Space Telescope*. After that, we could develop the technology further so as to reach faint objects and achieve still higher resolution. It took the radio astronomers 20 years to learn the art of very-long-baseline interferometry. It will probably take about as long for practitioners of the art of optical interferometry to catch up with them.

The fourth project involves active optics, that is, the use of optical interferometry to bring light nonrigid telescopes into exact focus. The different parts of telescope mirror surfaces would be held in exactly the right positions by servocontrols using feedback signals from interferometric sensors. Some efforts have been made to build an active-optics telescope on the ground with a quick enough response to compensate for the rapidly fluctuating distortions of the image caused by atmospheric turbulence. The same technology could be used far more effectively in space, where the distortions are caused by thermal relaxation of the telescope structure instead of by the atmosphere; thus, the time available to compensate for distortions is measured in minutes instead of milliseconds. If we work diligently at the techniques of active optics, we could soon have a poor man's *Space Telescope* with an aperture of one meter, which would be light and cheap enough to be produced and deployed in substantial numbers. We could then go step by step to bigger and more powerful optical arrays.

Of course, all of my suggestions have been connected with astronomy because that happens to be the area of my personal involvement with space science. Similar opportunities certainly exist for new departures in other areas such as solar physics, planetary exploration, and the study of interplanetary particles and fields, but I will not try to invent specific new missions in these areas. I am not claiming that the four astronomical projects which I listed are necessarily the best or the most important things for a space science program to do. I am claiming only that they illustrate a new style of operation, which could rescue all branches of the space science enterprise from the doldrums in which they are now stuck.

In astronomy the advantages of the new style are particularly clear. Optical astronomy would move, as radio astronomy already has, away from the old technology of single dishes and into the new technology of large arrays, phase-sensitive sensors, and sophisticated data-processing. *Space Telescope*, which is scheduled for launch in 1985, is 19th-century technology—a big rigid lump of glass. It represents the end of the old era rather than the beginning of the new, which will be flexible, both mechanically and psychologically. If we can begin now to explore in a modest way the technology of flexible optical arrays, we should arrive at the year 2007 with a variety of instruments surpassing *Space Telescope* in power and versatility as decisively as the radio-astronomical arrays of today surpass the radio-telescopes of the 1950s.

I would like to end my discussion with three additional examples of longer range technological initiatives, which may help to open new opportunities for the space scientists of the future. The first is an idea that has been talked about at the Jet Propulsion Laboratory, where it is called the "microspacecraft." The enormous advances in data-processing technology during the last 20 years have been a result of continued miniaturization of circuitry. The idea of the microspacecraft is to miniaturize space sensors and navigation and communications systems so that the whole apparatus is reduced in scale like a modern pocket calculator. There seems to be no law of physics that says that a high-performance exploratory spacecraft such as *Voyager* must weigh a ton. We might be able to build vehicles to do the same job in the one-kilogram weight-class.

A second technological initiative is a serious effort to exploit solar sails as a cheap and convenient means of transportation around the solar system, at least in the zone of the inner planets and the asteroid belt. Solar sailing has never seemed practical to the managers of NASA because the sails required to conduct interesting missions are too big. Roughly speaking, a one-ton payload requires a square kilometer of sail to drive it, and a square kilometer is an uncomfortably large size for the first experiments in packaging and deploying sails. Nobody wants to be the first astronaut to get tangled up in a square kilometer of sail. But the development of solar sails would be a far more manageable proposition if it went hand-in-hand with the development of microspacecraft. A one-kilogram microspacecraft would go nicely with a 30–meter-square sail, and a 30–meter square is a reasonable size to experiment with. The problems of sail management and payload miniaturization will be solved more easily together than separately.

My last technological initiative also fits harmoniously with the development of microspacecraft and solar sails. I call it the "space butterfly." It would exploit for the purposes of space science the biological technology that allows a humble caterpillar to wrap itself up in a chrysalis and emerge three weeks later transformed into a shimmering beauty of legs, antennae, and wings. Within the next 25 years this biological technology will likely be fully understood and available for us to copy. So we may be able to think of the microspacecraft of the year 2007 not as a structure of metal, glass, and silicon, but as a living creature, fed on earth like a caterpillar, launched into space as a chrysalis, and metamorphosing there like a butterfly. Once in space it will sprout wings in the shape of solar sails, thus neatly solving the sail-deployment problem. It will grow telescopic eyes to see where it is going, gossamer-fine antennae to receive and transmit radio signals, long springy legs to land and walk on the smaller asteroids, chemical sensors to taste the asteroidal minerals and the solar wind, and electric current-generating organs to orient its wings in the interplanetary magnetic field. A high-quality brain will enable it to coordinate its activities, navigate to its destination, and report its observations back to earth.

I am sorry. I claimed that I would not be imaginative, and now I am off again into the wild blue yonder. One last thought I leave with you. We may not have space butterflies by the year 2007, but it is a good bet that we shall have something equally new and strange, if only we turn our backs to the dead past and keep our eyes open for the opportunities which are beckoning us into the 21st century.

Comments

HENDRIK C. VAN DE HULST
Leyden Observatory
Leyden, The Netherlands

Freeman Dyson's inspiring paper provides an optimistic, daring vision of the prospects for science in space. It is my task to complement rather than to criticize what he has written.

I come from Europe, and some of my comments may reflect that fact. Europe is a continent of many languages and many cultures. The mixture of languages can be annoying but it also gives us an advantage. Far more often than the average American we have experienced situations when perfectly understandable words are not understood simply because they do not belong to the listener's vocabulary. This conditions us for the similarly blank faces that regularly occur in joint work among people from different disciplines. Such situations and the misunderstandings that may arise from them can always be overcome by patient efforts.

We have had the opportunity to test the severity of the language barrier in many real cooperative undertakings, not merely exchanges of ideas or the making of arrangements to coordinate otherwise independent measurements. A real collaboration is one in which work to produce the hardware and software of a project is parceled out among partners in such a way that the failure of one part means a complete failure of the project as a whole. The European gamma-ray satellite COS-B was such a real collaboration. Some of our American friends predicted that COS-B would never make it because it involved people from six institutes in four countries and speaking five native languages. Yet COS-B, launched in 1975 and retired only this spring after six and one-half years of excellent service, showed that—with luck—it can be done.

Europe spends considerably less money on space science than the United States does, in spite of comparable gross national products (GNP). The precise ratio of space science expenditures to GNP is hard to give because of the very different accounting systems used in the various countries. But whatever the actual figure, the difference is substantial. European scientists regret this, for it reduces the number of opportunities and makes proper planning more difficult. But it is a political fact, and I will not speculate on its causes.

HIPPARCOS

The encouraging words Dyson devoted to the European *Hipparcos* project pleased me. I say *encouragement* rather than *praise*, because a satellite should not be praised before it is launched and has performed. The European Space Agency selected this project from a number of design studies in March 1980, and we have good hopes that it will indeed have the impact on astronomy forecast by Dyson. It is not such a small project, however. It is estimated that the cost to completion will be on the order of 200-million dollars.

Because *Space Telescope* will have an extremely good astrometric capability, some people have wondered if *Hipparcos* does not represent an unnecessary duplication. This is not the case. If we compare stars in the sky with fiduciary marks on a table-wide construction drawing, then the capability of *Space Telescope* is to study the configuration of points that can be viewed together in the field of one magnifying glass. In contrast, *Hipparcos* can measure the mutual position of points across the table and thereby determine whether or not the drawing sheet is wrinkled or distorted. Indeed, minute positional distortions across the sky, mostly arising from climatic influences, form a notorious problem in astronomy. That is the problem that will be addressed by *Hipparcos*.

LOOKING TO THE FUTURE

Predicting the future is impossible. Works of art representing some future scene always carry a very clear signature of the time in which they were made and hardly any of the time they are supposed to represent.

Just for curiosity I recently reread a number of predictions on the future of space science that my colleagues and I wrote in the early 1960s. A mixture of admiration, fun, and some shame was the result. An admirable paper written by Herb Friedman[1] in 1965, when x-ray astronomy was only three years old, predicts that by 1985 that new kind of astronomy would be flourishing along with radio astronomy and optical astronomy. He also predicted that the x-ray observation of "hypothetical objects such a neutron stars" would become important. Well, it certainly has, and well before 1985.

In the same paper, Friedman expressed the expectation that in gamma-ray astronomy "clear prospects exist for the development of pictorial instruments utilizing emulsions or spark chambers, and the electronics for rapid analysis of large numbers of events with the necessary selectivity." This is a surprisingly accurate characterization of the *COS-B* satellite, especially the selective analysis of large numbers of events, which is literally true.[2]

Distinguishing and recording a true cosmic gamma photon among the events that trigger the various counters making up the total instrument is like trying to find a needle in a small haystack: a search for one from among 3×10^7. The detailed numbers are about like this:

• the outer guard counter, which discriminates against charged particles, triggers about 100,000 times per second;
• the combination of counters, which forms the "telescope," sends the OK signal to fire the spark chamber about 100,000 times per week; and
• a cosmic photon occurs among the events registered by the spark chamber about 100,000 times per year.

Typical gamma-ray sources in our galaxy, fortunately, but not predictably, have fluxes 50 times stronger than the one quantum per year per square-centimenter estimated by Friedman in his 1965 paper. The Vela source is even 400 times stronger. Friedman added the opinion that "the challenge is great enough to justify the most elaborate efforts." We were nevertheless relieved that the greater yield has brought down the price per *COS-B* gamma photon from the 10 thousands of dollars once feared to a more comfortable 100 dollars per photon.

MAN ON THE MOON

In the euphoria after *Sputnik I*, it was not uncommon to find books or articles offering space colonization as a solution to the rapid increase in the world's population. This irresponsible optimism has now subsided, although it does persist in some science fiction. Travel to the nearby moon seemed less fantastic, and many scientists expected this would become common in the near future. In his 1965 paper, Friedman foresaw that in 1985 "live astronomers may feel quite at home on the moon." He also speculated that x-ray star-rise and x-ray star-set would enable the moon-based astronomers to obtain accurate positions with modest equipment, and that the lunar base would house neutrino telescopes.

The step down from this vision to the facts and expectations of the present is more than a matter of a delay of a few decades, but I do not have the expert knowledge required for a deeper analysis. Personally, in the early sixties I did not expect to live to see man on the moon. I was wrong. The "giant step for mankind" *was* made. My doubts were based on the thought that it would be too expensive, and in that I may have been closer to the truth. Frankly, I do not now expect to live to see Freeman Dyson's butterflies unfolding their wings in space. I hope I am wrong again.

THE DYNAMICS OF PLANNING

In his implicit criticism of some past choices, Dyson addressed the general question of how a space science program can responsibly be planned. First, this responsibility cannot be dodged. Even in situations in which the scientist formally is only advisor and the power of decision lies with someone else, it is the scientist's heavy responsibility to give the right advice. All of my colleagues, in this country and abroad, take this responsibility seriously.

Secondly, the process of choosing a few projects to be realized from many proposals can be structured in various ways. The European Space Agency (ESA) has had success with starting from a long list of promising ideas, paring them to a shorter list for mission definition studies, then to an even shorter list of phase A studies. Finally, one is chosen as the next project. In this process of gradual choice, the initial stages are governed mostly by criteria of scientific importance. In the later stages, technical and financial factors add their weight to the decision. The eliminated projects drop out and have little chance of revival in a subsequent round, although the possibility is not specifically excluded. NASA relies more on the setting of priorities; delays of a number of years are more common, but dropouts are fewer. Both systems have advantages, and they probably will evolve toward each other.

Thirdly, making explicit the criteria used in the selection process has been an elusive goal. Basic and accidental considerations mix freely together. Philip Morrison [see Concluding Remarks, this volume] once made an analysis of the criteria used by the Space Science Board that has been very helpful to us.[3] Recently I also sketched how we have struggled with the definition of criteria in Europe.[4]

Fourth, it is impossible to play safe. There was considerable commotion last year about NASA's decision not to send a spacecraft to comet Halley. ESA made the positive decision to send a fly-by mission with the code name *GIOTTO*. It takes courage to take such a gamble. In fact, not having the capability of slowing the spacecraft during encounter means that the majority of the instruments aboard will perform the most interesting part of their task in 10 minutes, which means a year's money of the ESA science program spent (or lost) in that short a period of time.

Fifth, planning a joint ESA-NASA project involves the idiosyncracies of both organizations, including their different hierarchies and fiscal years. In spite of some mishaps in the past, I believe that the effort of coordination, including the realization of an occasional joint project, should continue.

Having made these five remarks, I will add the hope that in some time to come someone will be able to write a scholarly book about planning dynamics in space science. As in fluid dynamics, the author should introduce certain dimensionless numbers, like the Mach number and Reynold's number. May I suggest the following for this new field:

FR fruition ratio (some may read "frustration ratio")—the ratio of the number of projects executed to the number of projects through phase A planning.

P/C the ratio of project duration to career duration; this usually is about one-third (10 years/30 years). The fact that it has come so close to one has repercussions that are not yet well understood. The classical proverb *"ars longa, vita brevis"* assumes that in any endeavor of interest this ratio is much larger than one. The PI principle (PI = principal investigator) so popular with NASA assumes the opposite.

FTST the fraction of time spent thinking. This number is not specific for space science. If it gets too low, one may enter a different regime in which responsible decisions are no longer guaranteed.

PPratio the ratio of paper weight to payload weight (one of my colleagues insisted that this should be added). Evidently this ratio should remain much smaller than one.

SPACE TELESCOPE

First, *Space Telescope* is a joint project with unequal partners. An arrangement has been made between NASA and ESA whereby one out of every seven persons working on it will be a European, and one out of every seven dollars spent on it will come from a European pocket. The cooperation is excellent. A large telescope in space has been a prime wish of both American and European astronomers from the very start. Twenty years ago, in the Great Hall of the American National Academy of Sciences, I heard an inspired lecture by Lyman Spitzer on the large space telescope. *Copernicus* (*OAO-2*) and the International Ultraviolet Explorer (*IUE*), though they are both small and carry equipment that is now old-fashioned, have amply demonstrated the great potential of telescopes in space. Yet Dyson does not have much sympathy for *Space Telescope*, because it does not fit in the rule of "small is beautiful."

Being deeply involved in one part of the *Space Telescope* project, I have had related feelings; they have taken the form of worries. The telescope and its instruments are of enormous complexity, and so is the system of ground operations. How can the hundreds of people in industry and consultant firms, in NASA centers, in university departments, and in the Space Telescope Science Institute, who are preparing the systems and the sub-sub-subsystems, ever be motivated enough to make *Space Telescope* a success?

The Bible says that the tower of Babel, reaching into the heavens, was a flop because the builders ended up not understanding each other. There have been plenty of small signs showing the onset of such a danger with *Space Telescope*, but such dangers are present in any large project. I doubt personally whether the shield of micro-management that NASA puts up against this danger gives final protection. But the added devotion and good sense of the scores of scientists working on the project make me confident that *Space Telescope* can be a success. Probably this is what Harvey Brooks meant [see Section 1, p. 12, of this volume] when he spoke of "a unique blend of hierarchy and collegiality" as the basis for NASA's achievements.

A big project in space is by no means automatically a "brute-force" project. It may—and *Space Telescope* is a good example—employ in its minute details the same ingenuity that is characteristic of small projects.

LIMITS

Since the start of space research we have known that the sky is not the limit. But what in practice sets the limit to what we can do in space science?

The following answer sounds like a joke but must be examined seriously: It is good that there is not enough money, for something has to set the limit, and it would be very bad if there were not enough ideas. My friends do not like this formulation because it is pretentious and might give finance officers the wrong idea. Yet it does describe the situation in the European space science program, and presumably also in the present American program. The situation probably was different in the heyday of the Apollo program, when the bottleneck may have been the ability to train and manage enough people in time to meet the big goal that was set. It sounds simple— taking limited budgets as a common ground for planning, especially given the rather violent differences in attitude that may arise. Unfortunately, this is not the time to examine such questions in depth.

NOTES

1. H. Friedman, 1965, "The Next 20 Years of Space Science," *Astronautics and Aeronautics* 3(11): 40–47.
2. G. F. Bignami et al., 1975, "The *COS-B* Experiment for Gamma-Ray Astronomy," *Space Science Instrumentation* 1: 245–268; B. N. Swanenburg et al., 1981, "Second *COS-B* Catalog of High-Energy Gamma-Ray Sources," *Astrophys. J.* 243: 169–173; H. A. Mayer-Hasselwander et al., 1982, "Large-Scale Distribution of Galactic Gamma Radiation Observed by *COS-B*," *Astronomy and Astrophysics* 105: 164–175.
3. P. Morrison, 1975, "Rationale for Assignment of Priorities," ch. 4 (pp. 21–25) in *Opportunities and Choices in Space Science, 1974*, Space Science Board, National Academy of Sciences, Washington, D.C.
4. H. C. van de Hulst, 1983, "Planning Space Science," *ESA Journal*, April.

Comments

GERALD J. WASSERBURG
Geology and Geophysics
California Institute of Technology

Freeman Dyson has emphasized the virtues of a very limited set of activities, rather than present a balanced assessment of the remarkable and diverse achievements of space science. He has made some rather strong value judgments, which may not have a sound basis. I will try to deflect some of the things he has said and reflect on others. My approach will be to present some general comments, some questions, and some prognoses.

In 1492 Columbus discovered America, an event that we recently celebrated. About that time one of my heroes, Francois Rabelais, was born. He was a great scholar and intellect of his century, who barely escaped being burned at the stake. (I hope I can do the same!) Two of his works are *Gargantua* and *Pantagruel*. There is a section in these books devoted to the hero's explorations, which led him and his crew to the kingdom of the quintessence called Entelechy. This was ruled by a queen, who encouraged and supported numerous savants. The group of adventurers was, in fact, saved by residents of Entelechy, in particular one savant from the University of Paris. After recovering from their mishap, they were given, at the queen's request, a guided tour intended to exhibit the wonders of her empire and the technologies developed there. At one place, the guide took them to a hill from which they looked down upon a broad green grassy field where medieval scholars—wearing the pointed caps symbolic of their trade—were occupied. On inspection, they saw that these individuals, all carrying measuring rods, were running around, apparently randomly. A closer look showed that they were measuring the jumps of frogs, a large number of which populated the field. Well, the leader asked, what were they doing? And why? The guide explained that these were great savants and that it was their principal interest to study the lengths and directions of frog jumps. But what is the use of this endeavor? asked the leader. Well, the guide explained, these are some of the greatest and most creative scholars in Entelechy; they are the source of the methods that keep our land bountiful and prosperous in times of peace, aid us to be prepared in times of danger, and provide us with the means of victory in time of war.[1]

This story, written four and one-half centuries ago, shows that things have not changed much. The justification of the scientific enterprise is still embedded in the total human and social enterprise. The cogency of arguments brought forward in support of science is not always clear, but the growth of technology together with science continues to bear a complex but real relation to society.

About the frogs: I do not mind Freeman Dyson feeling a greater personal excitement over the jumps his frog makes as compared to the jumps my frog makes, but I cannot accept his statements about the intrinsic value of his frog versus my frog. In other words, his frog—he is, after all, a theoretical physicist—covers about two-thirds of the history of the universe; my frog covers only the most recent third. But mine includes the gathering of galactic debris made previously, the injection of matter from one or two supernova explosions, the formation of our star, the sun, from a local interstellar gas and dust cloud, the formation of the planets and the comets, the evolution of the sun and the planetary bodies, and the local formation and evolution of life, plus some hints as to what could happen next. My frog covers just our corner of the universe, and is not the whole story. But it is one of profound scientific interest, which may be pursued with dignity and vigorous scholarship. He may be tempted to recommend funding on a 2:1 basis, using the time-scales as a factor; but the available and immediate information from the local universe certainly demands immediate attention, which might be reason for making it 1:1 or even 1:2.

Now, the title of this section is "Science and Space," but I believe it should really be "Science and Exploration in Space." Historically, science and exploration have been strongly coupled. The charge of the Apollo program was to put a man on the moon and return him safely. Science was a minor issue. It is therefore unreasonable to demand that the scientific return justify the cost of the Apollo explorations, but rather that it provide a most exciting and substantial complement to the main venture.

The Apollo explorations carried some scientific experiments with them, and they returned lunar samples, which permitted us to study the evolution and structure of another known "planet" for the first time. The results have been most rewarding. About 10 years ago I wrote an article, titled "The Moon and Six Pence of Science," which is a bit of an overestimate of the cost of Apollo science. Given that we were already going to the moon, this sixpence was a beautiful coin of intellectual and scientific value.

However, we should not be too arrogant; the exploration and the science should be put into some perspective. In 1970 I was part of the U.S. delegation to the international Committee on Space Research (COSPAR) meeting in Leningrad. I was to present the *Apollo 11* scientific results. The audience consisted predominantly of scientists; the meeting hall was in an ex-Czarist palace. As I walked to the podium I noticed that the hall was full, and the aisles were becoming full. There was no room left either to sit or stand in that great hall. My talk was well received, and there was a large ovation when I finished, although I did notice that the audience seemed slightly restless. I then suggested to the next speaker that he was much in my debt for attracting so large an audience. His name was Neil Armstrong; so much for science with and without exploration.

Focusing once again on science in space, I note that Freeman Dyson has indeed brought forward a list of important science and science management issues. These include the need to preserve flexibility, to provide room and opportunity for new ideas and new technologies, to have available a wide variety of instruments and missions to take advantage of opportunities and new ideas, to reconcile the current time-scale of about 10 years for most spaceflight ventures with the more normal time-scale for scientific achievement, to have missions that are not just a one-shot affair, to have follow-on missions, and, lastly, to achieve a balance between large and small missions. I have tried to clarify Dyson's statements in order to convert them to a list of provocative issues, rather than a series of outrageous statements. I will not argue the merits of the *Space Telescope*, or the failure of *Copernicus*, or his claims for what he believes *Hipparcos* will achieve as compared to the four-meter telescope at Kitt Peak. Although we should pursue these matters, it seems to me that, at the present juncture, the science community has a more serious problem, which I will now bring forward.

Some more recent history pertaining to planetary sciences and exploration will serve as an example of where we are and where we are going. The situation is similar in several fields. The bountiful harvest of scientific results that we just reaped from the *Voyager* missions is the result of a program first approved in fiscal year (FY) 1972. The last new start in this field was six budget-cycles ago. Since FY-1974 (that is, almost 10 years ago) the program has proceeded downhill. This decline followed the commitment in 1972, made final in 1974, to eliminate expendable launch vehicles (namely, the Titan Centaur), which meant that the launch capability for deep-space exploration no longer existed. Last December it appeared that there would be a full close-out or extreme shrinkage of activities in aeronautics and applications. Private individuals like John Simpson and James Van Allen made great efforts to protect some science.

Today, circumstances hint at being a little better. Congress recently took some action to support a substantial upper stage for the shuttle, a capability that is urgently needed if we are to proceed with planetary exploration. There is a glimmer of light at what we hope is the end of the tunnel, but it is not a beacon. There may be a mission, but there is no program.

More generally, last month some colleagues in space physics, infrared astronomy, cosmic ray physics, and planetary sciences asked me some questions at lunch. Where is the United States in space science and where is it going? Why is there not a better assessment of U.S. and non-U.S. capabilities for use in establishing our plans? Where are the plans and the opportunities to fulfill them? Can we establish an environment that is compatible with consistent planning and with a consistent level of effort directed toward real goals? What and where is the involvement of universities in space science (in both doing and training)? Should we try to bring students and young people into space science? How can we maintain developments in science and new technology in universities? What is the future of both flight and nonflight space science in the United States? How can scientific experiments be designed and developed as an integral and intrinsic part of space missions (including shuttle and space-station payloads) and not as incidental add-ons? Where is science in NASA; what is the science advisory structure; and why is there no chief scientist?

Now there are some answers. There is a chief scientist—Frank McDonald; perhaps he is the glimmer of light. Perhaps it is the explicit inclusion of space science in the President's National Space Policy statement that is the glimmer. We had better move in that direction and hope that it will lead us toward an enlightened future.

The activities outlined in the National Space Policy recently presented by President Reagan established six basic goals. The first is to strengthen the security of the United States; the remaining five lie predominantly in the civilian sector. The policy statement clearly reaffirms the National Aeronautics and Space Act of 1958, which enunciates the NASA charter for activities in the civilian sector.

The growing need for using satellites in the area of national security is both obvious and necessary. In terms of commercial uses, particularly for communications satellites, there must be an ongoing and increased capability to loft these important payloads on schedule into the proper orbits. But what about the scientific and exploration part of the civilian space program? Surely, in the absence of a positive commitment to long-term objectives, the civilian space program will be subsumed in the military part, or else NASA will be reduced just to running a transportation system.

Many of the civilian activities will atrophy unless space policy is brought into balance by an active recognition that the civilian program is a vital ingredient of U.S. national policy. To me, this means that there must be full cognizance of the five remaining goals in the space policy, which are in balance with the first precept, the national security goal.

To manage the science part will require a strategy for space science and exploration. In some areas strategies have already been submitted by the National Academy of Sciences. These must be brought forward and applied at the White House and in Congress in order to establish substantial objectives on both the long and short time-scale. There must be a visible commitment to the missions necessary to achieve these goals, not simply as individual missions but as part of an ongoing program. This would permit the continuous development of a library of instruments, which could both grow and improve along the lines suggested by Freeman Dyson.

Given such an ongoing program, scientific management and planning must be an intrinsic part of the enterprise. This will require that the scientific community be willing to participate in all phases of planning, engineering, and decision making (as was the case in the *Voyager* and *Space Telescope* programs). On the other hand, NASA management must honestly solicit such serious participation, and not treat it as the incidental and formal addition of a high priest of science simply to bless an expedition that is all ready to depart.

I do not think that we can afford to come nearer to the foreclosure of space science than we already have. We learn from history, preferably other peoples' histories. The "China syndrome" is a great threat, but it is not the burnthrough of a reactor core; it is the cutting of lines of learning, doing, researching, and exploring in a modern society. We cannot withstand a 15-year withdrawal of active scientific development and training, nor can we withstand even a 5-year withdrawal of this type. We have already been injured by a simple lull of a few years, and that has hurt enough.

Our engineering faculties need substantial enhancement and support, and our science faculties need rejuvenation with young blood. These young people will need opportunities to do something. All of the science and engineering faculties in our educational system must go back to work creating new technologies and seeking new ideas, not writing a host of new proposals that will not be funded and that lead to work that will not be done.

To me this symposium celebrates the destruction of old myths and the creation of a new one. In every language there are allusions referring to impossible attainments like "asking for the moon" or "reaching for the stars." We have asked for and gotten the moon;

many of us have personally held some of her intimate parts in our gloved hands. To state that some enterprise has "as much chance as the man on the moon" is a vanishing allusion. These word-myths have been dispelled. Now there is a new myth: Mankind can achieve anything. Undoubtedly, that myth will also be dispelled; but for the time being, I would like us to try to live up to it.

NOTES

1. During the preparation for my commentary, I failed to restudy the works of Rabelais. As a result, I falsely represented the hopping creatures as frogs. They are fleas. Rather than rewrite the paper, I have left the frogs alone. The correct representation of the events in Entelechy may be found in the *Histories of Gargantua and Pantagruel* by Francois Rabelais (chapter 22 of the Fifth Book published ca. 1547). The English translation by J. M. Cohen (Penguin Books, 1972) is as follows: "Others were making a virtue of necessity, and it seemed a fine and proper job to me. Others were picking their teeth while fasting, a form of alchemy which helped very little to fill the close-stools. Others were carefully measuring fleahops in a long garden. This practice, they assured me, was more than necessary for the government of kingdoms, the conduct of wars, and the administration of republics. They claimed that Socrates, the first man to have brought philosophy down from heaven to earth, and the first to have transformed it from an idle trifling into a useful and profitable pursuit—that Socrates had spent his time measuring the hops of fleas, as Aristophanes the quint-essential testifies."

Discussion

FREEMAN DYSON   In almost all respects I agree with what Gerald Wasserburg has written, and I certainly do not claim that my frog is any better than his. It just happens to be the frog I am familiar with. I am deeply concerned with the dropping of planetary missions in the last five years, and I hope that we will move ahead and do more with the planets. I am saying only that we ought to try to do it a little bit differently.

DICK PRESTON (Star Foundation)   Do you think we will have a future in space if we do not do something about our science education? Is there any way that you as scientists can return to help us in the classroom?

GERALD WASSERBURG   As I wrote in my remarks, I consider the general status of science education in this country to represent the true China syndrome. It is not the meltdown of the core of a nuclear reactor; it is 15 years of withdrawal from active participation in scientific development and the training of young people. As I also wrote, we have already suffered a lull of a few years. As a nation we certainly cannot suffer for 15 years. Much has to be done to vitalize our engineering faculties, which need substantial enhancement and support; our science faculties, which need young blood and rejuvenation and the opportunity for young people to do and achieve things; and our educational system, which needs to permit creative research as well as teaching and the seeking of new ideas. Writing should be done, but not the production nationwide of 10,000 proposals a year for experiments and ideas that cannot in fact be explored. This situation must be addressed by the country, and undoubtedly it will be turned around as soon as people hear the message. But they will not hear the message from testimony like that delivered to the House Committee on Science and Technology in October 1982. That testimony unrealistically indicated that things were sort of all right, and that was rather disturbing.

FREEMAN DYSON   I have the impression that kids in the schools keep a sharp eye on the job market. If there are jobs, they will get the necessary training. It is not something you can push them into; rather, it is something they have to reach for. If they see that society is supporting this sort of activity, they will come forward and do it. That is what I see in my own family, and I think it is generally true. So, I think it is fine to try to improve the level of the schools, but one has to deal with the job market first, and then the schools will probably provide what is needed.

HENDRIK VAN DE HULST   I have no special knowledge of what is happening in this country, but my immediate reaction is to want to broaden the question. It is not science education that is wrong; people simply need more and better education.

CHARLES VERHAREN (Howard University)   I liked the comment, "Great science must involve surprises," and I wonder if that remark is borne out in the history of science.

GERALD HOLTON   There is almost a complete overlap between great science and surprises. The sentiment I keep encountering in studying the early modern period, the early 20th century, whether it is Planck, Heisenburg, Bohr, or Einstein, is: "We were driven to despair, and out of that despair came a conclusion that we otherwise would not have been able to reach for." And it was a surprise to themselves as well as to everyone around that these heroic acts really worked out. Of course, for every one of those that did work out, there were also bad surprises that also affected the progress of science. Certainly we can measure the quality by the degree of deviation from previous trajectories.

FREEMAN DYSON   One illuminating example is in Gerald Wasser-burg's own field. There was a magnificent program to bring back the moon rocks and study them on earth. Of course, to do that, NASA funded a lot of geophysicists and chemists and supplied them with very good instrumentation. Then, just through the goodness of the Lord, a big rock was thrown down from the sky and landed in Mexico, the Allende meteorite. It turned out, in fact, to be richer in isotopic anomalies, in all sorts of really exciting new information about the distribution of isotopes in the cosmos, than any data that came out of the moon rocks. Of course, the instruments were there, so if we had not had the moon program, we could not have done so much with the Allende meteorite. That is, I think, a very fine example of the way science really works; if we had planned to study meteorites, we would have done it quite differently. Fortunately, everything worked out for the best.

CHARLES GOODRICH (Goddard Space Flight Center)   I have two ques-tions which probably reflect my bias in space physics, which is that of particles and fields. The first is whether one should really consider *Galileo* as a one-shot program, when it is really an extension of *Pioneers 10* and *11* and *Voyagers 1* and *2*. *Galileo* will provide us with a lot of follow-up information on the magnetosphere of Jupiter, which has turned out to be a very surprising and very interesting thing. In that sense, I do not consider *Galileo* as a one-shot program, but as a more sophisticated instrument, a logical extension of what has gone before.

   The other question is this: In my experience with the space pro-gram, both on *Voyager* and now in a theoretical program, NASA tends to be essentially an experimental organization. What has been discussed here is largely centered on new missions; what sort of instruments can we send into space? My experience has been that NASA has underfunded and underemphasized the analysis of the data we have, putting all of its emphasis on acquiring new data. Should we rethink that issue a bit?

GERALD WASSERBURG   The space agency does not live by flight alone.

MORRIS HORNIK (George Washington University)   There seems to be a sentiment in space science that, until every last datum has been properly analyzed and placed into some kind of framework, the science is not complete. I do not think the history of science bears that out. If I am not mistaken, some specimens Darwin collected in three years on the *Beagle* have yet to be curated. Again, there was a French academy expedition halfway around the world prior to Newton's work. The important thing to remember about that expedition is not that the French did it or the number of papers that were written on the data collected, but that it gave Newton the numbers he needed for his own work, in his own study to prove the moon theorem and establish the laws of gravitation. Good science does not necessarily come from the amount of data or the level of analysis of the data. But how do we bring the results of space science as achieved today into contact with enough good minds actually to derive their full scientific value?

HENDRIK VAN DE HULST   If the data-gathering has been very costly, we ought to use the data. On the other hand, in all data and in all experimentation, certain features stand out and are important, and the rest can be forgotten.

DAVID BATCHELOR (Goddard Space Flight Center)   The scientific output of a number of NASA's missions has been evaluated today; one is conspicuous by its absence: *Skylab*. I think *Skylab* should be mentioned because observations of the sun made on *Skylab* led to several great advances in solar physics and provided a lot of stimulus to those working in the field. The significance of manned flight for space science has been downplayed by the members of the panel. What do they know of the discoveries of *Skylab*, and what do they think about the applicability of men in space doing science?

GERALD WASSERBURG   None of us could really cover anything but a brief view, so omission should not be considered an indication that we think something is not important. The real question that you raise, however, has indeed been a matter of substantial discussion: It has been manifest that there are an enormous number of activities that instruments can perform more cheaply and better than humans can. That is a fact agreed to by almost everybody that has contributed to this symposium. The question is whether or not, in the totality of the space enterprise, there is a substantial role for human beings. This has been raised for discussion several times; but it is a difficult question, and there is no simple answer.

SECTION 4

Conclusion

Concluding Remarks

PHILIP MORRISON
Massachusetts Institute of Technology

PERCEPTIONS AND CONTEXT

I remember one thing best of all about *Sputnik I*, 25 years ago. My immediate response was to rush to my high-frequency receiver, my shortwave set, and listen most of the night trying to find the 20–megacycle carrier frequency coming from the satellite overhead. I succeeded a few times. Around the world hundreds of thousands of persons must have done that. Among them were two clever people at the Applied Physics Laboratory, who were led to the invention of the first space navigation system by that experience. They could, in fact, measure Doppler shifts, which led them to a method of obtaining clear positional references. The situation is a little similar, perhaps, to the discovery of x-rays. Once x-rays were announced in 1896, they were generated and studied in laboratories around the world, and further discoveries were made within a week. Many, many people were caught up at a single time.

It was the broadcast of that physical signal the world over that lent unity to *Sputnik I*. Of course, the choice of high frequency for that particular piece of not very important telemetry, but decisive public relations, was crucial. Had it been done on VHF, only experts would have heard it; but transmitting at 20.003 megahertz, I think, was great—a shot heard round the world, not only in metaphor!

My role in this symposium is to lend further unity to the diverse, complex, and interesting points of view already presented. One way to do this is to catch at an epistemological thread, that is, to consider

129

how we have come to know that which we purport to know. A theorist is always rightly concerned about that.

Not long after its launch, I reflected upon *Sputnik* to write about it. It came to me then, something that is truly epistemological in nature, that the extraordinary virtue of the first artificial orbiting object was that it represented the practical, not merely the inferential, refutation of the Aristotelian cosmos. That is the world view in which that which is below belongs below; that which is above belongs above; that which is above is circular, perpetual, and shining; that which is below is transient, earthy, and marred; and the moon stands in between. It is not a bad cosmological view, a sort of first-order theory; and it is still held, I think, and perhaps rather firmly, by most people in the world—at least by a great many.

Even those who have been taught better, and who should know better, still hold onto Aristotle. They do not have a powerful epistemological conviction of how the contrasting results they have been given are known. They profess to believe that the earth spins and goes around the sun, but only because they know that is the received wisdom. When I have shown people the sun's image projected through a telescope, they have not cared at all about the sunspots I was revealing. They had all seen much better photographs of sunspots from much better telescopes. What fascinated them was that the image of the sun moved rather rapidly across the floor. The sun is moving in the sky, perceptually moving in the sky; this can be appreciated not just because at 11 o'clock it is in one place and at 3 o'clock it is in another. That is not the same thing at all. We can perceive velocities and motion directly, not merely changes in position. The eye and the mind can accept the new result at a quite immediate level of epistemological conviction, without the cold framework of geometry.

Ever since Nicholas of Cusa, as Gerald Holton told us, there has been reason to believe what Copernicus argued. The thought experiments, the extension of the world, and all of science come to bear. But if we simply reject those inferences and calculations, the numbers, the crosshairs, the clocks, and all the things that the physicists and astronomers have delighted in since those days, we can indeed deny the whole thing as a dreadful, if consistent, illusion. It is not easy to refute such a conviction; the true skeptic can remain quite content. But once we take a piece of metal from some Leningrad works—which certainly embodies a few designers' errors and quite a few machinists' marks and fingerprints where they passed the job on from one to another—and we carry it from its earthy home and throw it into the sky, it goes around the globe many, many times (a bit too small to be shining perhaps but circular and perpetual as anything). By adding momentum, increasing the quantity of momentum, we have turned an imperfect artifact into a celestial body.

All this was foreseen, of course. It was written about in very much this language by no less a novelist than Dostoevsky in *The Brothers Karamazov*. Nevertheless, the final action itself in some sense secures the enterprise of science. It proclaims that all those finicky inferences really mean something. This same sort of validation exists magnificently in the space age by the photographs of the earth's surface, which resemble nothing so much as the excellent maps we have all been raised on—maps expressing what people learned by triangulation of a non-Euclidian surface in a good many centuries of quantitative cartography. Now they are snapshots.

The match between the casually accepted photographs taken from geosynchronous orbit and the maps we all have long known is an extraordinary one. The sense of that reification, that making real to some form of perception, of what we could find only by measurement is extremely good. The whole chronicle of astronomy makes an extremely persuasive argument, a clinching argument, for any person who wants to adhere to the principles of scientific inference. But that is still not as good as the genuine perception of the new microplanet in the sky.

It may not even be enough just to hear the signal on the 20-meter band. It is really not enough to read of it in *The New York Times*. Something does not always happen if it is in the *Times*, and a lot of things happen that are not in the *Times*, too, as we all know. Every source is finite. However, once he actually sees the satellite, then the skeptic has a final quietus. I remember very well, having read *The New York Times* and believed the daily schedule published for the appearances of *Sputnik II* (or whatever it was), that I once recalled and made use of the data. I was very lucky. I was dining with several physicists near Philadelphia, who had not read the *Times*. I said: "Look, while we're just waiting for our next course to be served, let's walk out the front of the restaurant [I'd noticed the sunset orientation when we went in], and I'll bet we can see the thing come over." My watch was right enough, good luck in those pre-electronic days. We went outside, and in 20 seconds, no more, a sparkling point came majestically across the horizon, moving up the sky just after dusk on a clear winter evening! It was unforgettable to see the predictions of the Almanac realized, so to speak; that perception was very widespread.

Richard Lee, the Toronto ethnographer of the !Kung San, a group of bushmen of the Kalahari, told me of his experience. Once when he was there, sitting around the embers of the campfire under the velvet black skies of the Kalahari, he noticed that there were many satellite tracks visible overhead. (He is no astronomer, not even a space buff, but he knew a little something.) He could see the satellites moving among the stars. So one night he tried to ask his friends among the !Kung what they made of it, the unusual motion and so on. Though they laughed, they were a little reticent. They are a jovial people, and they knew he was kidding them a little bit, so they played the game. But he was so importunate that they said finally: "Okay, we'll tell you what we know about it, though we don't know very much. There are learned persons among us who know a lot, but we laymen here only know a little bit. What we know is that you made that, you outsiders. You put that up there. We know one more thing. We know when it goes from north to south it carries war with it from north to south. When it goes from south to north, it returns the war the other way." Now, this is perhaps a little apocalyptic; but as a crude first-order inference on the empty plains of the far Kalahari, I submit that that was a triumph of information collection applied to interpret direct experience.

A few remarks have been made already about the prehistory of *Sputnik*. That prehistory is worth mentioning. There is, of course, the trinity of grandfathers of the effort, the first three devoted workers bound to a dream of using space: Tsiolkovsky, Goddard, and Oberth. There are also three fathers from the next generation, the decade from 1925 to 1936 or so, when engineering investigation of these matters began. The best stories have been written about the group in Berlin, the German Rocket Society; there was a similar group in Leningrad, which rapidly acquired some support; and there was, of course, the foundation to fund von Kármán and the Caltech rocket research group, which began its rocket tests in the wastelands of the dry arroyo exactly on the spot where today the Jet Propulsion Laboratory spreads its big laboratories. There is a bronze plaque erected there to commemorate that event, just where a couple of graduate students went out to fire the little rockets. Such was the beginning in three countries; here we are, more or less 50 years from that beginning, marking a sustained effort in which the engineering continuity is strong.

In one way it is very hard to fix the actual beginning of the space age. I am perplexed, of course, because we all understand that there is a family of orbits. Keeping the eccentricity the same, if we raise one parameter alone, we will go from a closed to an open orbit. An open orbit is plainly in space, but what of the closed orbits followed by baseballs and stones, projectiles of all kinds? How are we to decide when we really are in space? Maybe the anniversary of the circular, perpetual orbit, the long-lasting earth orbit that has been chosen for this symposium, is a good one; but these are the doubts that come to one trying to decide. The long parabolas (of course, really truncated ellipses) were achieved earlier.

Naturally, there are several choices for an anniversary celebration, not only of what to commemorate and when to start counting, but also of how many years make a notable occasion. The introduction of the binary system into the otherwise overly austere metric system broadens those choices. We normally count by decades and centuries, but naturally we throw in a few 25s and 50s, just as in Italy one buys butter by the quarter- and half-kilo. Ten fingers are not everything, and the binary division of the world is also an extremely important one. Therefore, we have anniversaries by decades, and anniversaries of 25 and 50 years: 1957 is 25 years back. But we also have a significant 40th anniversary on almost the same day: October 3rd, 1942. On that day the A-4 (the *Aggregat-4*)—the cool laboratory name for what in grim war became the melodramatic V-2, 10 tons all-up, 7 tons of fuels and oxidizer, 25 tons of thrust—looped in its first 100-kilometer by 200-kilometer orbit over the Baltic Sea. That was the engineering beginning of orbiting spacecraft, for that first successful liquid-fuel, high-altitude rocket climbing well above the atmosphere is clearly linked with the present day.

As Bernard Schriever has mentioned—because of the German V-2, the implications of the development of control theory and practice and the development of powerful radiotelemetry—it became clear after the war (in 1946) that, with the same *Aggregat* and a perfectly plausible second stage, Peenemunde could have orbited a payload as early as 1942: 40 years ago. Indeed, the Japanese, with *Lambda-4*, orbited a payload with a rocket no more massive. It was simply one designed with the benefit of 25 years of engineering improvements.

In 1946 the Rand Corporation was formed. I was myself targeted as a recruit by two founders, the eminents Kohlbaum and Griggs. They invited me to join Rand, explaining that its purpose in the secret charter, not the one filed with the State Board of Corporations, was to wage intercontinental warfare by any and all means. They then thought that the missile with a nuclear warhead was the right weapon. When I objected, stating that I did not want to fight intercontinental warfare by any means whatever, they replied: "That's okay, we include the null case. You can come anyhow!" But I stayed away.

Furthermore, Convair, which became General Dynamics, made a substantial engineering study on the feasibility of intercontinental ballistic missiles in '46. This effort has been going on for some time.

There are two more anniversaries to recall, which are relevant, though they will not add to our cheer. The first is another decade anniversary, just 40 years, to be commemorated very seriously in Chicago. On December 2nd, 1942, "the Italian navigator entered the new world." Everyone knows the phrase means that Enrico Fermi and his group made the first sustained nuclear chain reaction: if the neutrons increase like $1 + k + k^2 + \ldots + \ldots$, as k approaches 1 it can be written $(1/1 - k)$, and that becomes a very big number. The number of neutrons they were making was only about 10^{18}, about the size of Avogadro's number; by now we on earth have made 10 or so powers of 10 more neutrons than that.

Moreover, on November 1, 1952—just 10 years from the first fission chain reaction, or 30 years ago—came the first open-ended thermonuclear explosion in the atolls of the Pacific. At once the somewhat quixotic V-2 was seized with powerful, possibly decisive, military importance. When the V-2 struck, the kinetic energy of the rocket body was greater than the explosive energy of the warhead. There was doubt whether it would be more effective to fire them loaded or empty. Given the 20 kiloton yield of a nuclear fission warhead, enthusiasts between 1946 and 1952 could argue that a rocket weapon was worthwhile. Given one megaton, the die was cast; the river was crossed. There was probably never enough opposition to impede such extraordinary systems, and they have now grown to thousands of orbital missiles and 10,000 warheads in each of the world's two great arsenals.

I point this out because it is perception that we are celebrating. We ought to celebrate the orbiting of little *Sputnik I* as the beginning of this our space age, but I think it only wise to recall that it had a context, less well-known at the time because it was under a cloak. It was not, of course, a very concealing cloak. The bulk in the jacket pocket of the IGY showed up very well to many who knew. But who knew? It was not made public to those who run while they read.

The famous defense advisory committee of John Von Neumann was established in 1953, well before *Sputnik*. Simon Ramo was himself a distinguished member of that small, powerful, farsighted committee. Another member, historian Herbert York, offered three reasons for the move to intercontinental ballistic missiles. The first was the presence of a new U.S. administration under General Eisenhower, which brought in new people with new ideas and a determination to do something new. The second was disturbing intelligence reports confirming what was already pretty clear: the USSR was making progress on ballistic missiles of oceanic range and had invested in them heavily. Third, and I think this was the decisive novelty, the thermonuclear explosion had made this weapon uniquely formidable. It seemed essential to at least study such systems. They were developed, I have to say, with a bang—that is to say, in phenomenal richness. There were many alternative approaches and contracts, and enormous industrial expansion rapidly occurred. The following list includes only those projects that involved formal government development contracts issued to contractors before we ever heard the news of *Sputnik*: Atlas, 1954; Thor, 1955; Titan, 1955; Jupiter, 1955; Polaris, 1956; Agena, 1956 (a very interesting one); and then the solid-fuel systems developed out of work done at Jet Propulsion Laboratory, the Minuteman, 1957. All preceded *Sputnik I*, including, as Schriever alluded to, one high-energy upper stage, the Agena, capable of orbiting a considerable payload into escape orbit.

All this was part of the large-scale Army, Air Force, and Navy development; the chief share was in the Air Force, but the others were not excluded. The complex origins of the IGY and the American concern with separating civilian and military activities—I think it a valid concern—delayed the orbiting of an American satellite. The first Soviet ICBM was tested in August 1957, a multiple-engined launch from the banks of the Jaxartes River in old Scythia. It was a large military booster with plenty of capability, and it was some months ahead of a U.S. ICBM. It was fairly clear that in a short time somewhere a lasting orbit would be entered instead of the long parabolic orbits of more direct military significance. It was Korolev himself, the principal designer and veteran of the old Leningrad days, who proposed the sputnik scheme as a dramatic stroke.

So there we were. The United States was plunged into a military orbital program on a big scale. By 1961, a new administration conceived the Space Olympics (as Ramo called it) for many reasons, not the least of them the extraordinary growth and influence of the aerospace industry. What an interesting way to use this tremendous new capability, a more benign aim than other plans surfacing in those years![1]

Wernher von Braun and his group at Peenemünde became the foundation, first at White Sands, and then at Huntsville, of a powerful rocket development team, something of an exotic element implanted into the American aerospace industry. That team is often the butt of a telling witticism that von Braun "aimed at the stars and hit London." That was, in fact, true for the German years.

I am not very sympathetic to the von Braun story, but, having read the documents carefully, I must also say that he was financed and directed to aim at Eastern Europe and, in fact, hit the moon. Of course, he succeeded in the original goal as well. Indeed, that sums up the history of the Huntsville enterprise nicely; having made the Jupiter IRBM, which lived only a short time in the field (much to the embarrassment of a later President of the United States), the Huntsville group was swept up in the NASA enterprise. By that time there were tremendous inputs from the vigorous U.S. aerospace industry. They had made all those intercontinental ballistic missiles in quantity; together the firms had gone far beyond the limited capabilities of even the very large enterprise at Huntsville. But Huntsville became the point of the civilian effort; Saturns were built eventually under Huntsville systems control, Saturn I, at least, in their very own works.

TWO DECADES OF CIVIL SPACE

Recent publications summarizing this 25th anniversary show the changing division of the U.S. budget for space activities over the years. Two government branches are important: the Department of Defense (DOD) and the National Aeronautics and Space Administration (NASA). For DOD, it is not the costly operational systems, the missile systems in place, that are counted, nor even their maintenance. It is the real, perpetual, circular orbiters—not the virtual, parabolic orbiters—alone that count. It is only the space segments and the costs of taking data from them whose totals are added. But all of that is now quite a big enterprise in the Air Force, and it is growing.

For the curves of annual space expenditures for DOD and NASA, I will discuss only the two intersections. The first was in 1961, when the annual NASA expenditures, civilian space expenditures, rose above the annual DOD expenditures. Of course, NASA was founded in 1958; a new small organization could not spend at a billion-dollar scale all at once. But it took only two or three years for them to acquire the know-how, and they succeeded brilliantly in 1961, with their total rising rapidly. In 1981 that intersection was achieved once again, but with NASA dropping below DOD. This fiscal year, and in many subsequent ones, NASA expenditures in space will be smaller than that of DOD. *The New York Times* of September 29 quotes Undersecretary Ulrich of the Air Force, projecting future military expenditures in space to rise to about twice NASA's amount in 1988. NASA will spend more or less the same dollar sum, losing real buying power as a result of inflation. In contrast, the DOD space budget will increase by about seven percent per year after inflation, about the same rate as the military budget as a whole. (Of course, this is only in the budget papers and has not yet received electoral or even congressional approval.) In a way, here is the end of a historical epoch, the priority of civilian over military goals, at least as expressed by one crude measure.

The partition—costly and imperfect, but clear—between civil and military, peaceful and warlike, open and secret, national and worldwide, has been one of the most admirable features of our American space enterprise. I look with dismay upon what will likely occur in the near future, with the breakdown of these barriers. It was the 1958 Space Act that created NASA as a civilian agency—open and devoted to international cooperation. All of those things, of course, are much more difficult to maintain under a military regime, for evident reasons of long standing.

This change is not without concern for science and scientists. I applaud the enthusiastic and eloquent account of *Uhuru* given by Freeman Dyson. I remember *Uhuru* very well; I heard the launch countdown from Malindi by radio, and I recall the great excitement as each new varying signal came in. I was lucky enough to live close to the place where the data were being taken, and it did reveal, or at least seal, the new fast-changing cosmos that we had not understood very well before. But here is an unpublished anecdote, a plausible and well-founded rumor, the accuracy of which I cannot fully guarantee.

The project called *Vela Hotel*, which provided a multiple-satellite, timed x-ray detection system, was set up in the mid-1960s. It called for keeping, on the average, a half-dozen or so small satellites in cislunar space. The Air Force administered it as a method of fulfilling President Kennedy's promise—made as a condition for United States ratification of the Partial Test Ban Treaty—that we would provide ourselves with those famous "unilateral means of verification" against any nuclear test explosions in space. Now, space is not the best place to test nuclear explosives secretly, so that danger was not great. But in those days of big budgets, we were willing to put money into such systems, and we did.

Of course, *Vela Hotel* recorded all kinds of x-ray sources and their time variations; but between studying the new system and being unwilling to disclose its capabilities to a possible violator, this material remained long unpublished. It was regularly examined with startling and enthusiastic inferences by the Los Alamos group, which received the data, at arms length as it were, at the end of a long chain from the Air Force's collection stations, quite without knowing exactly what it meant or exactly what was going on. But they could not publish. So for several years they alone knew there was an x-ray sky full of variations until *Uhuru* broke the ice; then it was clear what they had been seeing. Soon they themselves were able to publish their famous discovery of the gamma-ray bursts, which actually were first detected in the secret *Vela Hotel* system. (To be sure, this somewhat folkloric account is unconfirmed: better take it as a parable!)

The past Soviet space program was xenophobic and tightly controlled. Even attainment of orbit was denied public notice until all was secure and safe, and no firing was reported in live detail. The Soviets have moved toward a more open system of communication, both in public relations and within their own scientific community. They have also allowed foreign observers and foreign instrumentation. Now, even a French cosmonaut (known as a spationaut) flies in their system. Today the drift in the USA—with a minor secret mission mixed into a shuttle test—is toward a kind of convergence to an unpleasant mean between the two systems, our system being a little Sovietized and theirs being Americanized.

The next substantial increment in NASA activity will very probably be in the direction of space platforms, with manned space stations being placed in near-earth orbit in the decade ahead. Many people draw the plausible inference that the shuttle system itself has as its principal objective the handling of the multiple payloads required to assemble those substantial permanent objects. The platforms are well-planned. They have excellent attributes, but I am afraid that the principal motivation behind them, apart from a general economic effect in the NASA centers and in the aerospace industry, will be found in a speech by Chairman Leonid Brezhnev given in August 1982 to the Unispace Conference in Vienna. He said, in passing, "The future will see large orbital complexes with changing crews, permanently operational in space." That is only an undated prediction. On the other hand, it is a prediction toward which the Soviet technologists have moved steadily. They have a lot of practice; they have done a lot of things about replacing crews; and they have developed the fundamental technology quite well. I strongly suspect that our intelligence people firmly believe that the Soviet Union will indeed go through with that, and I expect we will guarantee it by starting some such program ourselves.

The program has its justifications; I am not saying the scheme is all wrong. I am saying is that this is a very different world from the one we were talking about during the days of NASA's budget peak. Dominated as it was by Apollo, NASA expenditures still fed space science, space applications, and deep-space exploration. The large space platform is mainly man-in-space, largely for national prestige.

There are utilitarian features. The studies offer a very interesting mixture of manned and unmanned platforms, as appropriate to mission and circumstances. I hope that becomes the case. There is a kind of physical edge that cuts between most military and most civilian missions, though that is not to say there is strict division. At least, in one viewgraph in the famous NASA style, we see a polar orbit for one space station and a near-equatorial orbit for another. The people that the military most want to look at live at high latitudes, and one had better not study them along all that slant range from low, equatorial orbit; a polar orbit surveys everything and is clearly more attractive. A polar platform could also operate a few earth resources satellites, just as an equatorial platform could operate astronomical satellites. Those satellites would be dependent upon the big platform, possibly even launched from it and certainly with data collection, visit, repair, and control functions there. Here is one principal source of debate about the direction of NASA's future. I know of no other sizeable program that has such strong apparent support in government at the present time.

ON THE DARKNESS OF SPACE

The study of our future, we who live transiently on the earth, and the work of the historians who try to appraise our past, is a business of looking at a complex of systems with many reverberation times, many relaxation times. Each event echoes for a longer or shorter time; some only for a very short time in piercing sounds that are soon gone. I wonder whether there is going to be any residuum of this entire affair so powerful as the one that was presciently forecast in 1948 by a compatriot of Arthur Clarke's—a similar seer and writer as well as a distinguished astronomer, Fred Hoyle. In a book on astronomy he published in 1950, Hoyle wrote, "Once a photograph of the earth taken from outside is available, . . . once let the sheer isolation of the Earth become plain . . ., a new idea as powerful as any in history will be let loose."[2] It took 18 or 20 years until NASA let that idea loose in real photography.

When was that first view available? That depends on our definition. Does a slightly cropped earth count? Does black-and-white count as much as color? What about an earth deep in shadow? In August 1966, *Lunar Orbiter I* took that marvelous distant photograph of earthrise in black-and-white above the dusty moon. It was followed by more *Lunar Orbiters*. Then came a DOD satellite called *DODGE* with geophysical tasks; in September 1966, probably the earliest of all, *DODGE* had genuine black-and-white photographs of the whole earth. But I have never seen them; as far as I know, no one outside of a limited circle has seen them. Then came *ATS-1* in January 1967, with black-and-white shots from synchronous orbit, and then *ATS-3* on November 10, 1967, with a full-color photo from synchronous orbit. That very picture of the full earth remains the most circulated photograph of all the archival treasures of NASA, which says something for the sense of importance that Hoyle felt. For many years, that photograph has been a kind of symbol for the environmental movement, with its evocation of the fragility, isolation, and uniqueness of our blue earth. And an eloquent image it is.

There are other sides. A man in California—author, publisher, and innovator—a clever man called Stewart Brand of Sausalito, decided, in an epiphany not altogether generated by his own endomorphins, that what the world needed most was the sort of photograph that Fred Hoyle wrote about. So he took himself from a California rooftop to the Valley, then to New York, then to Cambridge, Massachusetts, then to Washington, D.C. In each place this young man took a stand, placing himself now in front of the door of Columbia University, then in front of MIT, also in Harvard Square, and certainly in front of NASA headquarters. He stationed himself there to sell to all comers, at cost, plastic badges that asked in large black letters on white background: "WHY HAVE WE NOT YET SEEN A PHOTOGRAPH OF THE WHOLE EARTH?" Probably the missions to do that were not yet in place. But he likes to think that out of synchrony, or some other mystery, not necessarily simple causality, his single-handed enterprise bore rich fruit, convincing the scientific-technical community how valuable this task was. Sure enough, he did this in 1965; in 1966 the real thing appeared.

The situation in 1957 was similar. *Sputnik I* was the first realization of something ready to happen somewhere. The solvent of history was saturated for its crystallization. When I wrote my part of a physics textbook for high school students during the summer months of 1957, I included in the introductory chapter a statement that before long an artificial satellite would circle the earth, going at high speed, without fuel driving it. This remark was just to arouse interest in Newtonian physics. I had no close knowledge of intelligence reports of the Soviet ICBM or of any other hint; I had only the scientist's gossip and the newspaper reader's knowledge that this gem was ready to form in the solution.

That great pulse of civilian space activity, manned and lunar as it was, dragged along with it an enormous technology of brilliant and breathtaking power. Now we are to some extent the masters of space, the masters of orbital momentum. It arose out of the perception that something extraordinary had happened when people far away, whom we then regarded less than we ought to have by a good deal, produced that steady blip in the 20–megacycle band.

It is surely epistemological, perceptual in the broad sense, that we have arrived at our present state. It was this manifest and unarguable demonstration, this way of learning it, that made the difference. I have no doubt there were many cogent reports on how this was going to come, what would happen next, who was doing it, how much was being spent, and so on. We ourselves were all but ready; it turned out otherwise, a world crossroads of perception.

PERCEPTIONS OF THE FUTURE

There is now a general dissatisfaction with the state of the world. We cannot forever base the future on what the scientific prowess of this country, and a few other highly developed countries, holds in trust for the rest of world—scientific prowess that has brought world history to this point. We seek more than the rationale of a simple straightforward nationalism, of benefit for country. However fruitful that idea has been and remains in history, there must be a more lasting image of the unified blue planet somewhere along the path we follow. Even the most narrow-minded practitioner of statecraft would accede to the view that we do live on a single round planet, a fact now, through our technology, made manifest to all. It is no longer an abstract notion; we can see it.

When Leo Szilard tried to understand the $N \rightarrow 2N$ reaction in London on the eve of war in 1937–1938, he used an expensive radium source. He had paid for it as a personal investment, so that he could be a free researcher, go where he would with his little radium source and do nuclear experiments, that is, if someone would provide some of the other necessities—a laboratory room, essentially. He tried to do this at St. Bartholomew's Hospital, a famous London institution, which had a distinguished department of radiology. For one reason or another they said no. He lost his temper and explained that, though it might be inconvenient for them, it would be still more inconvenient if he were not allowed to do his experiments because the building in which he was making the argument would soon no longer be there. The tale is ironic for if he had not done the experiments elsewhere, perhaps the chain reaction would not have happened. (The room did, in fact, disappear in the blitz, but that was not due to any actions of Szilard or to his nuclear competitors in Germany.)

NASA headquarters, the National Academy of Sciences, and the National Air and Space Museum are still here, but they are all vulnerable. The awareness of the vulnerability of the memorials of the very creations we are commenting upon is, I think, not lost on anyone. The best way we can ensure not only a satisfactory expectation of more grand experiments in space, but, on a much wider scale, even the expectation of any kind of successful outcome to the enterprise of science and technology, is by paying great attention to the future we now face. For the first time in human history, many kinds of activities are moving into—if not in terms of geophysics, at least in terms of geography—the same domain inhabited by the human species as a whole.

Von Neumann himself said it 25 years ago: Weapons' strike and weapons' reach get bigger all the time, but the area of earth stays constant. This implies an asymptotic situation, which bodes no good; that much is patent. The natural introduction of war into space, within our grasp, as discussed by Bernard Schriever, is a most dangerous enterprise, and it should not be entered upon if we can possibly avoid it. We might avoid it. We might inhibit it by treaty, perhaps by the proposed antisatellite treaty. I think there is a very good prospect for that. In the first place, no substantial system of the kind has been tested; such tests would surely have been visible. In the second place, the leadership of each nuclear, spacefaring, orbit-dominating power gains by wide knowledge and swift communication, and stands only to lose if the other side at some moment or other can strike down its satellites. We might inhibit it at an early stage, and the fact that space war does not add to what is called in the jargon "stability" suggests hope that there might be such agreement. I cannot myself see the wide destruction of satellites or the disabling of working satellites in orbit, except in a situation tantamount to catastrophic war.

Satellites in peace are a major part of the information industry, the most rapidly growing of industrial activities. We cannot take the view that either the beaming down of information to a pliant mass of viewers or the exploitation and maintenance of intense economic contrasts on the earth's surface by means of the diverse economic advantages provided by access to orbit is an enduring solution to the human condition. However temporarily pleasing, advantageous, or satisfying of creative needs and drives such use of space might be, we can do better than that. I hope that the perception of one isolated planet—to be sure, with disparate cultures, disparate economies, disparate expectations of life to the widest degree—will come over time to signify more than the fluctuating utility that this or that new system promises. To some extent this has been demonstrated by our general technological history. We have a proliferation of devices, yet no single one has been so persistent as the general pursuit of technical means of modifying the environment, making a living, and extending the making of a living. First those abilities were centered in the Renaissance Mediterranean; then they spread to England, which put them on a thermal basis. Then they spread to Holland, France, America, and Germany; then Russia became involved; and now there is no question that Japan, out of the game only 120 years ago, is a full partner. Looking at the demography of the world, one cannot doubt that the center of gravity is going to shift still more (barring a catastrophe), slowly moving somewhere into East Asia, where human beings live in such great numbers.

This is looking ahead much farther than we can look, beyond sensible extrapolation. Nevertheless, it is an inference from history. It was a dream of the Enlightenment—older than the National Academy of Sciences by 75 or 100 years, but well known to the founding fathers of the republic—that science and the steady increment of knowledge and its technical application would bring good to all. We no longer find this easy to accept; indeed, the evidence against it is strong. But as a dream, it is not entirely lost; it remains part of the internationalism of our science and technology. It remains part of our profession that we stand linked in an intricate chain, connected to the ancient inventors of the zero and, for that matter, the domesticators of fire. We do not easily overlook that, nor can we omit the day-to-day fact that the world of science still tries to maintain a sense of internationalism and community. Open publication is still the Mertonian norm; it is the exception in science when something is not published.

It has indeed been a splendid time for the happy few, which includes us—the scientists lucky enough to be working in areas affected by these remarkable events, lucky enough to read of and understand them, lucky enough to debate with our colleagues and to make our own proposals, even if they do not all fly. It has been a good time, also, for the many more people in this country, and in other countries as well, not the 10,000 or so in the coterie of the happy few, but the technicians and the other people who work in the enormous social enterprises that stand behind every large-scale novelty. They run to 10- or 20-million Americans, persons who regard themselves as part of the events, even if only because a relative works among these wonders. They read about it, and they feel themselves part of a renewed Copernican view of the world, a general view of the world as a big blue marble. Of course, they are puzzled and disturbed; they see problems. We all know that; so do the scientists. Yet they, too, have shared this splendid time, and they augment the happy few. They are among the happy because they are one part in a thousand, or five parts in a thousand of the world's population. That's 20 million people, if I am generous. But all the rest have a very curious relationship to it all.

Of course, space technology is not everything. Much has been said in the context of space, for which it stands as surrogate because it is the showiest and in some ways the most sinister surrogate for all of science and technology in our present world—sinister because of its close relation to the deadliest of the weapons of destruction. Nuclear physics shares the same burden. We have poured upon this 1957 event quite a lot more, I think, than it merits by itself. Perhaps that is as it should be, because those who have contributed to this symposium saw it as it pervades our time. Hendrik van de Hulst pointed out that if you try to draw the future, when people look at it some years later, they recognize exactly the year that you drew it because you cannot get out of your time. It is even more striking than that. If you simply try to counterfeit the past, the typological experts can date the trials once time passes. If you look at imitations, they stand out in the museums like sore thumbs. They are Victorian; they are not Greek at all. But no one saw that when the whole world was Victorian. The signal was not distinguishable from the noise. Once the noise has disappeared, the unintended signal stands out.

So when we see the future poorly, we see ourselves; listen to and watch anxiously everything we do in science and technology, in uncertainty and in doubt, just as many watch and few produce the widely broadcast programs of our communications media. It remains unclear whether over the next 25 years we will judge that the end of the Aristotelian cosmos was worth its cost, still unbilled. I close, then, in a sense of doubt, but not as a pessimist. The future has the remarkable property that it is plastic. It lies in our hands, and it needs treatment. It needs vigorous, intelligent, devoted, exhausting concern. Unless we are concerned enough, just as the mushroom cloud and the sites of Alamogordo and Los Alamos lie under a historical uncertainty, so too the first demonstration that Aristotle had it wrong will remain of uncertain value. But it has brought with it one inspiring lesson: We too, not gods by any means, but men and women with imperfect hands, can hope to launch something circular, perpetual, and shining.

NOTES

1. I myself heard a distinguished civilian aide to the Secretary of the Air Force advocate the bold scheme of placing thermonuclear missiles into solar orbit—not into earth orbit or just lunar orbit, but solar orbit, because then you could hide them behind the sun. They would be very far away, very hard to detect at two astronomical units. They could come down—if you believed everything would work properly—upon the right targets, if all was precisely timed as they flew back over a year. This never caught on. It presented for me the first sign of a system that nobody wanted to buy, hardly even its proponents.
2. F. Hoyle, 1950, *The Nature of the Universe*, Harper and Brothers, New York, pp. 16–17.

Index

active optics, 103, 104
Advanced Research Projects Agency (ARPA), 29
Advent communications satellite proposal, 29
Agena upper stage rocket, 136
Air Force. *See* United States Air Force
airline navigation and traffic control, 56, 68, 69, 72-74, 81
Alamogordo, N.M., 147
Allende meteorite, 124
Almond, Gabriel, 33
antisatellite treaty, 144; weapons, 29, 40, 144
Apollo program, 5, 9-11, 12, 16-18, 27, 29, 30, 38, 49, 65, 94-96, 114, 117, 140
Applied Physics Laboratory, 129
Ariane, 23, 32, 50, 62, 74, 79, 81, 100
Arianespace, 23, 74
Aristotle, 85, 95, 130, 147
Armstrong, Neil, 49, 117
Arnold, Gen. Henry H. ("Hap"), 27, 28
asteroid mining, 81
astrometry, 100, 101, 108
AT&T, 65
Atlas missile, 29, 37, 136
Atomic Energy Commission, 12
atoms-for-peace program, 14

Baruch, Bernard, 8
Batchelor, David, 125
Blagonravov, Anatoli A., 47
Bohr, Niels, 123
Brand, Stewart, 141
Brezhnev, Leonid, 140

Brooks, Harvey, 4, 27-31, 33, 37, 38, 41, 113
Bruno, Giordano, 86
Burke, Bernard, 102

C³I (command, control, communications, and intelligence), 31, 40, 41
California Institute of Technology (Caltech), 49, 50, 89, 115, 133
Cernan, Eugene, 94
Chase Econometrics, 76, 77
Chevalier, Roger, 50, 72, 79-80, 81
Civil Aeronautics Board (CAB), 68
Clarke, Arthur C., 8, 25, 48, 141
Colombo, Guiseppe, 96
Columbia University, 5, 142
comets, 67, 80, 111, 116
communications, satellite, 8, 14, 18, 20, 22, 23, 29-31, 36, 40, 42, 43, 48, 54, 55, 57, 58, 64-67, 68, 74, 79, 119
Communications Satellite Corporation (COMSAT), 65
Copernicus, Nicolas, 86, 95, 130
Copernicus (OAO-2), 91-93, 99, 112, 118
COS-B, 109, 111
COSPAR, 89, 117
Crab Nebula, 95

Darwin, Charles, 125
Department of Defense (DOD), 137, 138, 141
Discoverer project, 29
DODGE military satellite, 141
Dostoevsky, F., 131

Dyson, Freeman J., 87, 88, 90, 107, 108, 110, 115-118, 120, 122-124, 138

early-warning satellite programs, 28, 29
Earth resources satellites, 14, 18, 20, 23, 56, 69-71, 75, 82, 140
Einstein, Albert, 95, 123
Einstein (HEAO-2), 96-98
Eisenhower, Dwight D., 14, 28, 34, 135
Etzioni, Amitai, 4, 5, 33, 37, 38, 42, 43
European Space Agency (ESA), 23, 24, 75, 79, 100, 108, 111, 112

Federal Aviation Administration (FAA), 68
Federal Communications Commission (FCC), 57, 68
Federation of American Scientists, 88
Fermi, Enrico, 134
Field, George, 39-41
Ford, Gerald R., 49
Friedman, Herbert, 95, 109, 110; Morris, 44

Gagarin, Yuri, 41
Galileo, Galilei, 86
Galileo Jupiter probe, 99-101, 124
Galois, Evariste, 93
gamma-ray astronomy, 109; bursts, 139
Gardeneir, John, 81
Gardner, Traymor, 29
Gargantua and Pantagruel, 115
Gemini project, 29, 35
General Accounting Office (GAO), 77
General Dynamics Corporation, 134
General Electric Company, 49
geostationary orbit, 8, 18, 22, 55, 57, 100, 131
German Rocket Society, 133
GIOTTO Halley's comet probe, 111
Goddard, Robert, 132
Goodrich, Charles, 124

Hale, Edward Everett, 79
Hall, R. Cargill, 8

Halley's comet, 80, 111
Harvard University, 4, 6, 50, 85, 88, 142
Harvard-Smithsonian Center for Astrophysics, 39
Headrick, D. R., 10
Heisenburg, Werner, 123
Hertzfeld, Henry, 77
Hinners, Noel W., 38, 39
Hipparcos, 99-103, 108, 118
Holton, Gerald, 85, 123, 130
Hornik, Morris, 25
Hoyle, Fred, 141, 142
Huntsville, Alabama, 136, 137

Institute for Defense Analysis, 50
Intelsat, 22, 23, 24, 64, 74
Intercontinental Ballistic Missiles (ICBMs), 4, 8, 28, 52, 65, 66, 68, 136, 142. *See also* Atlas, Minuteman, Titan
International Astronautical Federation (IAF), 50, 73, 75, 79
International Geophysical Year (IGY), 7, 47, 51, 135, 136
International Solar Polar Mission, 80
International Ultraviolet Explorer (IUE), 112
IT&T, 65

Jet Propulsion Laboratory, 96, 102, 104, 133, 136
Joels, Kerry M., 42, 43
Jupiter, 67, 80, 99, 124; IRBM, 136, 137

Kardashev, Nikolai, 102
Katyusha rocket, 47
Kennedy, John F., 5, 9, 10, 34, 42, 138
Kepler, Johannes, 85, 88
Killian, James R., 28
Korolev, Sergei, 136
Koyre, Alexander, 86
!Kung San, 132

Landsat, 23, 69-71, 75, 82. *See also* Earth resources satellites
laser systems, space, 31, 40

Lee, Richard, 132
Lerher, Daniel, 34
Logsdon, John, 3, 25, 37, 39, 41, 42, 47
Los Alamos, N.M., 139, 147
Lubell, Samuel, 34
lunar module, 49
Lunar Orbiter I, 141

man-in-space, 6, 12, 13, 15, 27, 31, 38, 39, 44, 74
Manned Orbiting Laboratory, 29
Mansfield, Edwin, 50, 76
manufacturing in space, 15, 16, 58, 73, 81
MARECS, 79
Mariner 10, 96
Mars, 61, 94, 96, 98, 99
Mathematica, 77, 78
McDonald, Frank, 119
Mercury, 94, 96; project, 29
Midas early-warning satellite program, 29
Miller, Jon, 37
Minnaert, M., 88
Minuteman ICBM, 10, 29, 37, 136
Massachusetts Institute of Technology (MIT), 102, 142
Morrison, Philip, 111, 114, 129

NASTRAN, 77
National Academy of Sciences, 47, 88, 89, 112, 114, 120, 144, 145
National Aeronautics and Space Act of 1958, 11, 12, 14, 19, 29, 31, 119, 138
National Aeronautics and Space Administration (NASA), 14, 15, 18, 28-30, 35, 39, 44, 61, 68, 71, 76-78, 92, 99, 105, 111-113, 119, 120, 124, 125, 137-140, 141, 142, 144
National Air and Space Museum (NASM) 38, 42, 144
National Defense Education Act, 9
National Institutes of Health, 12
National Research Council, 58
National Science Foundation, 12

national security, 4, 5, 28, 30-32, 39, 42, 57, 119
National Weather Service, 69
NATO, 38, 55
navigation, satellite, 14, 20, 21, 54, 79, 129
neutrino telescopes, 110
neutron stars, 109
New York Times, The, 47, 49, 131, 138
Newton, Isaac, 48, 86, 95, 125
Nicholas of Cusa, 86, 130
Nova Cygni, 93
nuclear war, 31, 40, 41, 54, 144

Oberth, Hermann, 132
Office of Space Science and Applications (NASA), 92
optical interferometry, 103
Office of Scientific Research and Development, 12

parking orbit. *See* geostationary orbit
Partial Test Ban Treaty, 138
particle beam weapons, 21, 31, 40
Phillips, Gen. Samuel, 30
Planck, Max, 123
Polaris missile, 10, 136
President's Science Advisory Committee (PSAC), 3, 4, 9, 25, 28
Preston, Bob, 102; Dick, 37, 38, 79, 122
pulsars, 21
Purcell, Edward, 88

Rabelais, F., 115
radio astronomy, 88, 92
Ramo, Simon, 49, 50, 51, 72-75, 76, 78, 81
Ramo-Wooldridge Company, 49
Rand Corporation, 7, 8, 28, 134
RCA, 65
Reagan, Ronald, 30, 119
reconnaissance satellites, 28, 30, 55, 60, 67
remote sensing. *See* Earth resources satellites

Russia. *See* Union of Soviet Socialist Republics

Salyut space stations, 44
Saturn, 80, 103
Saturn I, launch vehicle, 137
Schmitt, Harrison, 94
Schriever, Bernard A., 4, 27, 35, 39-42, 44, 134, 136, 144
Shepard, Alan B., 41
shuttle. *See* space shuttle
Simpson, John A., 118
Skylab, 96, 97, 125
Smithsonian Institution, 89
solar power, 58, 73; sails, 105
solar system exploration, 6, 15, 16, 18, 61, 74, 85, 96, 97
Soviet Union. *See* Union of Soviet Socialist Republics
Space Act. *See* National Aeronautics and Space Act of 1958
space colonization, 110
space race, 3, 51, 52
Space Science Board (NAS), 88, 111
space shuttle, 13, 15, 17-2, 23, 24, 30-32, 33, 44, 50, 61-63, 66, 71, 79, 81, 96, 99, 100, 118, 119, 139
space stations, 44, 63, 73, 79, 139, 140
Space Task Group of 1969, 16
Space Telescope, 98, 99, 101, 103, 104, 108, 112, 113, 118, 120
Space Telescope Science Institute, 113
Spacelab, 24, 79
Spitzer, Lyman, 92, 112
SPOT, 23, 74, 75
Sputnik I, 3, 8, 10, 27-30, 33, 34, 37, 42, 47-49, 51-53, 60, 86, 94, 101, 110, 129, 130, 135, 142; *II*, 131
Sullivan, Walter, 47
supernova explosions, 116
Szilard, Leo, 143

Thor missile, 136
Titan missile, 29, 37, 118, 136
TRW Inc., 29, 49-51

Tsiolkovsky, Constantin, 47, 132
twenty-one centimeter line, 88, 89

Uhuru (Explorer 42), 94, 96, 101, 138, 139
Union of Soviet Socialist Republics (USSR), 3, 7, 10, 14, 34, 35, 40, 43, 47, 51, 53, 63, 87, 102, 140
United Nations, 38
United States Air Force, 4, 7, 27-30, 35, 38, 39, 40, 44, 68, 136, 137-39; Systems Command, 5
United States Information Agency, 35

V-2 rocket, 7, 47, 133-35
Van Allen, James, 118
van de Hulst, Hendrik C., 88, 89, 107, 123, 125, 146
Vandenberg Air Force Base, 30
Vela Hotel, 138, 139
Venus, 67, 96
Verharen, Charles, 123
Verne, Jules, 6, 48
Viking Mars probe, 96-99, 136
very-long-baseline interferometry (VLBI), 102, 103
von Braun, Wernher, 136
von Karman, 27, 133
Von Neumann, 135, 144
Voyager Jupiter probes, 96, 97, 105, 118, 120, 124

Wasserburg, Gerald J., 89, 115, 122, 124, 125
weather satellites, 14, 55, 56, 69, 72, 73
Webb, James, 11, 35
Wells, H. G. , 6
Western Union, 65
White Sands, N.M., 136
Wilson, Charles, 28, 41
Wright brothers, 7

x-ray astronomy, 109

York, Herbert, 135